Protected Species and Biodiversity

Protected Species and Biodiversity
A GUIDE FOR PLANNERS AND ECOLOGISTS

Tim Reed

Pelagic Publishing | www.pelagicpublishing.com

First published in 2024 by
Pelagic Publishing
20–22 Wenlock Road
London N1 7GU, UK

www.pelagicpublishing.com

Protected Species and Biodiversity: A Guide for Planners and Ecologists

Copyright © 2024 Tim Reed

The moral rights of the author have been asserted by them in accordance with the Copyright, Designs and Patents Act 1988.

All rights reserved. Apart from short excerpts for use in research or for reviews, no part of this document may be printed or reproduced, stored in a retrieval system, or transmitted in any form or by any means, electronic, mechanical, photocopying, recording, now known or hereafter invented or otherwise without prior permission from the publisher.

https://doi.org/10.53061/MASF8870

A CIP record for this book is available from the British Library

ISBN 978-1-78427-501-3 Hbk
ISBN 978-1-78427-502-0 Pbk
ISBN 978-1-78427-503-7 ePub
ISBN 978-1-78427-504-4 PDF

Cover images: Natterer's Bat *Myotis nattereri* drinking in flight from a lily pond © Kim Taylor/naturepl.com; Great Crested Newt *Triturus cristatus* male in a garden pond at night, surrounded by Water Fleas *Daphnia pulex* © Nick Upton/naturepl.com; Traditional organic hay meadow with a profusion of wild flowers and grasses including Pyramidal Orchid *Anacamptis pyramidalis*, Common Spotted-orchid *Dactylorhiza fuchsii*, Rough Hawkbit *Leontodon hispidus* and Yellow-rattle *Rhinanthus minor* © Nick Upton/naturepl.com; European Badger *Meles meles* foraging in front of house at night © SCOTLAND: The Big Picture/naturepl.com

This book is dedicated to the many ecological consultants that do their best to provide good-quality data to local authority planners, and to the planners that struggle with often poor data provided by those consultants who should (and do) know better.

It is hoped that this guide will make the lives of planners just that little bit easier and help the protected species and biodiversity that are, in effect, in their gift.

Contents

Foreword ix
Preface and acknowledgements xii
List of abbreviations xiii

Introduction 1

A green and pleasant land? 1
Why do LPAs need help with protected species
when evaluating planning applications? 2
How to use this book 3

1. The Planning System, Protected Species and Biodiversity 5

1.1 The policy needs for biodiversity evaluation 5
1.2 What do LPAs say they want? 12
1.3 ALGE, ecology and planning 16

2. Guidance and Interpretation 26

2.1 NE guidance 2014–21 27
2.2 NED guidance 2022 32
2.3 The Partnership for Biodiversity in Planning 46

3. Getting Better Data to Planners: Consultants, Data Quality, and Constructing More Suitable Guidance 62

3.1 The British biodiversity standard BS 42020 63

4. The New 2022 Standing Advice: Turning Around and Moving Forwards 71

4.1 Establishing the facts 71
4.2 What does an LPA need to have? 73
4.3 What does the January 2022 SA expect of an LPA? 73
4.4 Moving forwards from the 2022 SA in 11 steps 75

5. Surveys for Protected Species: What the LPA Might Have Ordered — 87
    5.1 Headings: what to expect and why — 87
    5.2 Progressing — 93
    5.3 Protected species guidance for planners in tables — 94

6. Using the Data from Effective Protected Species and Other Surveys — 206
    6.1 Biodiversity Net Gain and species — 209
    6.2 Testing BNG — 212
    6.3 Delivering protected species into the future — 214

Appendix: Key elements for each species/species group — 216

References and further reading — 232
Index — 243

Foreword

This book is a guide to the use of data on biodiversity, especially protected species, in the face of development pressures and opportunities. It is written for ecologists, planners, landowners, residents, advisers, indeed anyone with an interest, concern or obsession regarding development. It is an invaluable source book, and candidly points to problems and limitations in current guidance – and considers how to deal with these.

In the UK, as in many world states, governments have declared climate and nature crises. The seemingly relentless losses of semi-natural habitats and wildlife, attested in a succession of *State of Nature* reports, point to environmental attrition. In the UK, Mark Cocker captures the mood in *Our Place: Can We Save Britain's Wildlife Before It Is Too Late?* (2018), describing the first *State of Nature* (2013) report as a 'remarkable, and remarkably troubling, document, not least because it involved the collaboration of twenty-five different environmental organisations'. Indeed, 'the bottom-line figures in the *State of Nature* report measure not just the scale of loss, but the rate of decline. They don't indicate the bottom of a curve: they chart the direction of an arrow. It means that, however bad things are, they will get worse without major change.'

Globally, in *A Life on Our Planet: My Witness Statement and a Vision for the Future* (2020), Sir David Attenborough, aged 94 at the time of writing his book, refers to the Chernobyl reactor explosion in April 1986 being regarded by many as the costliest environmental catastrophe in history. He chides: 'Sadly this isn't true. Something else has been unfolding, everywhere, across the globe, barely noticeable from day to day for much of the last century. This too is happening as the result of bad planning and human error.' He is of course referring to 'the true tragedy of our time: the spiralling decline of our planet's biodiversity'. Chernobyl, the climate emergency and the biodiversity crisis are all down to bad planning and human errors!

This book could have as its subtitle 'how to avoid bad planning and human errors', for these have contributed massively to biodiversity loss. It is published at a pivotal time for development planning and implementation across Britain and Ireland. On 21 July 2024, the UK Government's Deputy Prime Minister and the Secretary of State for the Department for Environment, Food and Rural Affairs (Defra) wrote to the environmental NGOs setting out the ambition of a proposed Planning and Infrastructure Bill for England. They explained: 'When it comes to the planning system's role in providing the nature and housing we need, we know that the status quo is not working. Environmental assessments and case-by-case negotiations of mitigation and compensation measures often slow down the delivery of much-needed housing and infrastructure. Meanwhile, the condition of our environment, and even our most important habitats and species, has declined over a sustained period. This is a lose-lose situation, for our economy, the public and for the natural environment … Our vision for an improved planning system will require government to work in partnership with civil society, communities and business.'

In Scotland, the *National Planning Framework 4* (NPF4), published in February 2023, sets out spatial principles, regional priorities, national developments and national planning policy. Two years earlier, the Welsh Government published *Future Wales: the national plan 2040*, setting the direction for development there until 2040. In Northern Ireland, under the aegis of the *Regional Development Strategy (RDS) 2035*, Local Development Plans are at various stages of adoption or development across each council area. In Ireland, the renewed *National Development Plan 2021–2030* published in October 2021 covers 'priority solutions to strengthen housing, climate ambitions, transport, healthcare and jobs growth in every region and economic renewal for the decade ahead'.

All of this points to the complex national, regional and local governance around development planning, signalling changes ahead and the need for good quality data to underpin everything. Focusing mainly on England, but with wider applicability, this book is well timed to advise professionals and other interested parties on how planning works in relation to biodiversity, and in particular planning applications. Local planning authorities (LPAs), statutory conservation agencies, notably Natural England (NE), Defra, sponsors, ecological consultants, and a plethora of interested parties all deal with planning development and outcomes. This book takes the reader through an eleven-step process covering the planning process as it applies to protected species. The main block of the text provides tables for taxonomic groups and individual protected species, detailing survey methods, limitations and ways of collecting suitable data for planning submissions. These tables can be used by planners, developers and interested parties working on a planning case. Having so much material accessible as a single source helps greatly in plugging a gap in existing texts available to planners across the UK.

The final chapter covering so-called Biodiversity Net Gain (BNG) is a salutary pointer to the care that is needed in using available data to assess impacts, and why reliable data affect both the use of mitigation measures and the separate process of estimating BNG. Importantly, the BNG metric cannot be used to deliver protected species; it is a proxy for the negative impacts on habitats arising from a development, and it then calculates how much new or restored habitat (and types) is required to deliver sufficient 'net gain'. The approach is in its infancy, hotly debated, does not cover protected species *per se*, and is not the panacea for accommodating development pressures that some attest. Considerable amounts of guidance are being devised on this, with the Chartered Institute of Ecology and Environmental Management (CIEEM) offering regular updates on its website, under the heading 'Biodiversity Net Gain – Principles and Guidance for UK Construction and Developments'.

FOREWORD

Tim is admirably qualified to write this book as an active Fellow of CIEEM, and member of its Professional Standards Committee. I first met him in 1978 on the Isle of Rum NNR whilst he was undertaking his DPhil research based at the Edward Grey Institute, University of Oxford. Tim was studying land bird population ecology in the Hebrides, and one of his study sites was the small Harris woodland plantation close to the tiny lodge on the southwestern edge of the diamond-shaped island. I was there as an assistant leader on a Schools Hebridean Society expedition, and Tim joined us for a week, teaching expedition members (mainly school students) the intricacies of netting and ringing birds, as well as imparting his wide-ranging experience of bird ecology. As an undergraduate student, I marvelled at Tim's meticulous approach to fieldwork, and his critical assessments of what the data were revealing and further detailed work needed. Reflecting on that time, it is no surprise that Tim has gone on to be a leader in guiding ecologists and planners alike on sustaining biodiversity.

After completing his DPhil in 1981, Tim joined the then Nature Conservancy Council (NCC), leading field teams surveying upland and peatland birds, with much of his work focused on, and indeed much later contributing to, the inscription of the peatland Flow Country as a World Heritage Site in 2024. Following promotion to Head of Ornithology in the NCC, Tim held a Nuffield and Leverhulme Travelling Fellowship studying North American conservation management and audit systems, before returning to the UK to join the newly formed Joint Nature Conservation Committee in 1991. There, he headed up the development of UK-wide biodiversity recording and monitoring standards for the statutory and NGO conservation sectors. In 1998, Tim left the public sector for consultancy, working around the world with Flora and Fauna International, based in Cambridge, primarily as a corporate biodiversity analyst advising on biodiversity and ecosystem services' risks and opportunities. In the UK, he has served as an expert witness at wind farm public inquiries, focusing on submissions' data quality, and alas their unsuitability for use in support of far too many poorly framed planning applications.

Writing and publishing widely, Tim has emerged as a leading expert adviser on planning and the underlying evidence-base vital to successful sustainable developments. This guide is instructive, and timed just when development pressures will become greater than at any time in recent decades. We may be entering the era of win-win or of greater losses of biodiversity. As an eternal optimist, I hope we are closer to the former, and if we are then this important source book will have played its part.

Professor Des Thompson FCIEEM, FRSE

Edinburgh

Preface: who is this book for?

This book is primarily for the planners in local authorities, as they search for meaning and reliability in the ecological data on protected species accompanying planning applications submitted to them. It is also for developers – small and large – and their ecologists, who need to make their submissions as simple, clear and unequivocal as possible. It is for householders and landholders who need guidance on what they should commission and expect. And it is for neighbours or others who puzzle over how to make sense of what is put on council websites. Most importantly, it is for the protected species and biodiversity of the UK; if planning submissions are good, well documented and unequivocally establish proper baselines and allow the assessment of impacts, then there is the possibility that all may not be lost… and some may even be gained.

All we need to know is what surveys (desk and field) actually took place, what they mean, what they don't mean, how planners can confidently evaluate the data they are provided with and understand what these mean in space and time. If all of those involved in this understand their roles, and produce what is needed, clearly and explicitly, then we might all be winners. If not, then it is business as usual and we all, eventually, lose. If so, we risk developing a 'brown and unpleasant land' (Helm 2019).

Acknowledgements

This guide owes its genesis to discussions with a number of consultants deeply unhappy at the quality of data provided by others to local planning authorities. Their suggestion that someone should tackle the subject were well meant, but I now know why they stood back and let others do it.

Thanks go to a range of people, including Mark Avery for allowing space in his blog for me to tentatively air my concerns; to Nigel Massen for reading that blog and encouraging me to produce a first draft; to Mike Alexander and Darren Frost for reading several drafts and making critically constructive comments; to nameless reviewers who helped improve the draft further; to owners of copyright for allowing me to quote from their works, and for permission to use figures; and to Lissie Wright for withstanding the demands of writing when easier options might have prevailed. All errors and omissions are solely mine.

Abbreviations

ALGE	Association of Local Government Ecologists
BNG	Biodiversity Net Gain
BS	British Standard
BSG	Bird Survey Guidance
BSI	British Standards Institute
CIEEM	Chartered Institute of Ecological and Environmental Management
DEFRA	Department for Environment and Rural Affairs
EcIA	Ecological Impact Assessment
EIA	Environmental Impact Assessment
EPS	European Protected Species
LPA	local planning authority
NE	Natural England
NED	Natural England and DEFRA
NERC	Natural Environment and Rural Communities
NBN	National Biodiversity Network
NPPF	National Policy Planning Framework
NS	NatureScot
PBP	Partnership for Biodiversity in Planning
PEA	Preliminary Ecological Assessment
PPP	Prepare a Planning Proposal for Protected Species
PSD	Protected Species and Development: advice for local planning authorities
PSSA	Protected Species Standing Advice
SA	standing advice
SGN	Species Guidance Note
VC	validation checklists
WAC	Wildlife Assessment Check
WeBS	Wetland Bird Survey

Introduction

A green and pleasant land?

Over time, the British, and especially the English, landscape has come to be viewed through some sort of soft-focus lens as a green idyll (Rackham 1986). Add in the bucolic playing of brass bands, and there is a dreamy sense of yesteryear when all was well and plentiful.

After Blake (1808) set out his vision of England's green and pleasant land, the power of this image grew, blossoming with its incorporation in the music of the hymn *Jerusalem* (Parry 1916; Elgar 1922). The apparently verdant English landscape has now achieved almost mythical status.

Like most myths, there is a risk that pricking the balloon will deflate the rhetoric and replace it with a harsher reality. For decades, there have been reports of species declines and losses across the UK, gradual impoverishment of the urban and rural landscapes, and the hollowing out of their biodiversity (Cocker 2018; Carrington 2019; Environment Agency 2022). The periodic State of Nature Reports (Hayhow et al. 2019; Burns et al. 2023) paint a progressively less and less rosy picture for the UK, detailing reductions and geographic retractions across a wide range of species groups from insects through to birds and orchids. With approximately 40% of species in moderate or steep decline – and this from an already heavily depleted background state (Sheail 1998) – Britain is regularly described as being 'among the most nature-depleted countries in the world' (DEFRA 2023). Even a group as resolutely apolitical, and scientific, as the UK's mammal society (Mammal Society 2023) remarked:

> One in four of our native mammals is threatened with extinction, and many others are in decline. With Britain now recognised as one of the most nature-depleted countries in the world, urgent action is needed.

That is not the picture of a pleasant land, and perhaps not even a green one (Helm 2019).

Many of these biodiversity losses have occurred as agriculture has changed, and with it the mosaic landscape which supported a wide range of formerly common species (Cocker 2018; RTPI 2019; Hayhow et al. 2019; Burns et al. 2023). We are still losing the remaining ancient woodland, meadows, hedgerows and other habitats at an alarming rate. Gone too are huge numbers of insects, birds and other plants and animals reliant on them for survival (Hayhow et al. 2019; Burns et al. 2023). These include moths declining by 88%, ground beetles by 72% and butterflies by 76% (Carrington 2019). Between 1970 and 2010, 75% of the 20 million farmland birds were lost. Between 2002 and 2013 more than half of all of UK species declined (Tree 2018). As the UK's Natural History Museum put it in 2020: the UK has led the world in destroying its natural environment (NHM 2020). That is obviously nothing to be proud of.

Many of the changes have seen the breakup of once continuous blocks of habitat into much smaller fragments or the repeated salami-slicing of residual blocks of habitats (Cocker 2018). As a result, each supports disproportionately fewer individuals of a species, or numbers of species, than before, and many of these are common and the same. With it comes an increased risk of extinction too: a future faced by 14% of the UK's species in 2019 (Hayhow et al. 2019) and 16% in 2023 (Burns et al. 2023).

In spite of the English Government's 25-year Plan for A Green Future for England (Gove 2018) and the Environment Bill (2021), the state of the English and the wider devolved administrations' biodiversity resources are not something to celebrate. One in 6 species in Wales is at risk of extinction; this is more than 1 in 10 in Scotland and more than 1 in 10 in Northern Ireland.[1]

Changes that have occurred in the farmed and upland areas of the UK have mostly been outside of the developmental planning system. Under a wide range of incentive payments, these show no obvious sign of stopping (Helm 2019). As a result, there is increased pressure on areas of land within the local authority planning system to help balance and stem the losses and keep connectivity between blocks of suitable habitats in urban and rural areas.

With an ever-growing demand for housing, a steadily increasing population and central and devolved governments committed to economic growth – but with a stated policy of having no net impact on the environment, as habitats are altered or lost (DEFRA 2016), and requiring a net gain of habitat for any new developments (NPPF (HMG 2021)); Scottish Government 2018; Gove 2018; Environment Bill 2021) – the pressures on natural resources continue to rise (Cocker 2018; Horner and Davidson 2020). Whether a balance can be achieved or not is uncertain.

These losses call into question the continuing capacity of the natural environment, and the remaining depleted ecosystems, to provide clean air, water, fertile soils and somewhere for everyday enjoyment ('ecosystem services') as our natural capital continues to be shredded (Helm 2019; https://www.naturalcapitalinitiative.org.uk; Environment Agency 2022). No matter what term is used, it is critical that all potential impacts that affect these resources are understood: individually and cumulatively. Many of these impacts will be assessed under the umbrella of the planning system.

Why do LPAs need help with protected species when evaluating planning applications?

For those without the necessary skills, knowledge (Oxford 2012; ALGE 2016; Snell and Oxford 2022) or time to research the literature (much of it is either too complex or general) when reviewing ecological and protected species data, local planning authority (LPA) officers rely on the quality of submissions and the probity of data and claims accompanying an application. If they are fortunate, and few are (HoL 2018), contracted-out ecological advice may be available for an LPA officer.

A series of reviews by the Association of Local Government Ecologists (ALGE 2013, 2016, 2020; Snell and Oxford 2022; Boulton et al. 2021) showed that the data normally provided to LPAs are inconsistent at best. Boulton et al. (2021) call for better data to accompany planning applications. In the decade since being first reported by ALGE (Snell and Oxford 2022), little has changed. As a result, the need for a single reference resource – however imperfect – that helps LPA staff weed out the good from the bad in ecological submissions is only likely to grow.

The role of this primer is to help identify those gaps for protected species – which form the core of ecological planning submission data – and to provide practitioners with a simple basis with which to make an informed, reliable appraisal of what is coming over their desk. It should also allow them to understand and evaluate the advice sometimes offered from elsewhere within the authority and to provide a serious reference point when queried – as they will be – by councillors and developers seeking to add pressure to push a proposal through.

How to use this book

For those wishing to understand the background to statutory protected species material in planning applications, its origins, LPA needs and the problems associated with delivering it at a local authority level, Chapter 1 is a short introduction. Chapter 2 looks at the iterations of Natural England and DEFRA (NED) guidance produced by Natural England (NE), and some of the practical problems associated with NE standing advice. These are developed in Chapter 4. Alternative approaches offered by several groups to the material in the standing advice are reviewed and evaluated, including a better, statutory, option operated in Scotland.

For any potential supplier of protected species material to planners, and for planners themselves, the brief Chapter 3 is important. It summarises the key components of the British Biodiversity Standard BS 42020 (2013). The Biodiversity Standard underpins the LPA's expectations of data quality, availability and transparency in all planning applications, as well as why these are required in planning submissions.

Chapter 4 looks in more detail at the revised 2022 Natural England and DEFRA standing advice and some of the serious problems that remain, as well as how to deal with these in a positive and practical way. Eleven steps are given which, if applied, will provide what planners, sponsors and ecologists should expect when protected species (and non-protected species) are covered in a planning application. Understanding these steps helps when using the core tables of Chapter 5.

Chapter 5 is the main section of the guide and fills almost half of the space in the book. It can be used by itself but would benefit from readers looking at Chapter 3 and at the 11 steps given in Chapter 4. The first part of Chapter 5 introduces a suite of headings that should be covered for all protected species addressed in planning applications. These are worked through in detail in individual species-specific tables. Each table should be used by planners and applicants alike where the species occur to provide a common ground for the methods to be used, the recognition of limitations and the suitability of material that accompanies a planning application.

Chapter 6 looks briefly at the limited relationship between protected species and biodiversity net gain processes, as well as some of the challenges that remain as a result.

1. The Planning System, Protected Species and Biodiversity

When a planning application thumps onto a desk – or more usually makes a high-pitched *ping!* as it arrives in the computer in-tray – it contains a whole load of details. Among the issues listed in the set of cover sheets that normally accompany a planning application will be queries about protected species, biodiversity and geological conservation. This usually comes after flood risk and before foul sewage. It may well have tick boxes about protected species, the proximity of designated sites and waterbodies within 200 m. It is not exhaustive, and a naïve reader might be forgiven for thinking that biodiversity and protected species are therefore no big deal.

If biodiversity and protected species literally only tick two boxes, why should planners worry about biodiversity and ecosystem services (or natural capital as it has been renamed – Natural Capital Initiative https://www.naturalcapitalinitiative.org.uk) in planning applications?

This is because LPA planners are the first line of defence in terms of assessing and agreeing, or rejecting, planning proposals. They are the cornerstone of the edifices that are regional and national levels of biodiversity policy delivery and reporting, and they have a legal duty to cover biodiversity (NERC 2006; NPPF 2021; Environment Act 2021), which includes protected species.

1.1 The policy needs for biodiversity evaluation

As far back as 2006, as the updated National Planning Policy Framework (NPPF) (HMG 2021) notes, the Natural Environment and Rural Communities (NERC) Act of 2006 (HMG 2006) placed a clear duty on LPAs to deliver biodiversity and ecosystems conservation at the local level on behalf of central government. As NE's (2022) guidance on planning puts it:

> Section 40 of the Natural Environment and Rural Communities Act 2006, which places a duty on all public authorities in England and Wales to have regard, in the exercise of their functions, to the purpose of conserving biodiversity. A key purpose of this duty is to embed consideration of biodiversity

as an integral part of policy and decision-making throughout the public sector, which should be seeking to make a significant contribution to the achievement of the commitments made by government in its Biodiversity 2020 strategy.

As well as the sense of duty, the other important term used here is 'embed'. Embedding in this context means that everything should be seen through the prism of conserving biodiversity. Clearly, that requires a bit of skill and knowledge. Understanding what is needed, and how to do it, was apparently straightforward, and hopefully suitable guidance was available from Natural England (2022):

> Guidance on statutory obligations concerning designated sites and protected species is published separately because its application is wider than planning and links are provided to external guidance.

That Natural England (2014 and 2022) guidance is covered in Chapter 2 and in detail in the Appendix. Little of it is suitable for LPA needs.

In addition to the 2006 NERC Act (HMG 2006), there is a clear duty under Section 17 of the Crime and Disorder Act 1998 to prevent wildlife crime, as outlined in the 1981 Wildlife and Countryside Act and as amended. Any incidents are reported to the police and their wildlife crime officers.

> Without prejudice to any other obligation imposed on it, it shall be the duty of each authority to which this section applies to exercise its various functions with due regard to the likely effect of the exercise of those functions on, and the need to do all that it reasonably can to prevent:
> (1) crime and disorder in its area (including anti-social and other behaviour adversely affecting the local environment).

This includes biodiversity and protected species. Having raised expectations of a fully embedded appraisal of all planning applications by LPAs, NE sought to temper the risk of too much zeal – and attempted to put the genie back in the bottle – by stating that:

> Local planning authorities should take a pragmatic approach – the aim should be to fulfil statutory obligations in a way that minimises delays and burdens.

The result is a bit of a Catch-22: cover in all aspects, but not so deeply that it causes problems to developers. It brings out an early use of the concept of burdens: something onerous that might, in some way, reduce the speed of the otherwise exemplary unencumbered developer. It of course assumes that LPAs know how to pragmatically minimise delays and burdens while still fulfilling statutory obligations – another Catch-22.

Just what are the statutory obligations that might prove burdensome to developers? In England they are covered in detail in the NPPF (HMG 2021). In chapter 15 of the NPPF it states that planning should contribute to conserving nature and securing net gains for biodiversity. Each of the devolved administrations has its own analogous statements or policies. The Northern Ireland Strategic Planning Policy Statement (2015, paragraph 6.171) recognises the statutory duty of LPAs to conserve biodiversity. The Scottish Planning Policy (2014) does much the same, and in paragraph 194 it recognises the vital services provided by ecosystems. In Wales, the term is resilience. Planning Policy Wales Edition 10 of 2018 links this duty to section 6 of the Environment (Wales) Act of 2016 by:

- maintaining and enhancing biodiversity;
- not causing significant loss of habitats or populations of species;
- providing a net benefit to biodiversity;
- maintaining and enhancing green infrastructure.

In addition, all 4 countries have their own biodiversity strategy documents, none of which were updated after 2020, although Scotland (Scottish Government 2020) created statements of intent.

As part of their statutory duties, LPAs must also protect designated sites (Table 1.1). These vary from local nature reserves, designated by LPAs, through to sites of national and international importance, some of which carry multiple designations.

In addition, a 2015 decision by the Planning Inspectorate confirmed the duty of LPAs to include biodiversity when assessing permitted developments, while certain cases also confirmed the need to consider species and habitats covered by the amended 2017 Conservation of Habitats and Species Regulations (RTPI 2019).

What this all means is that LPAs cannot ignore protected species, biodiversity and ecosystem services in their daily approach and, as the RTPI (2019) noted, they would risk long and expensive legal cases if they did. As will be seen later, there is a difference between not ignoring and recognising protected species and biodiversity issues with the submission of suitable methods and data in a planning application.

1.1.1 LPAs and biodiversity: the strategic approach

Clearly, a reactive piecemeal approach will not work for LPAs. A decade or more ago, the Association of Local Government Ecologists (ALGE) set out in a flow chart the basic planning process (Fig. 1.1).

Figure 1.1 underlines the need for good quality, up-to-date data and the importance of clear guidance at the LPA level to help potential planning applicants. At the same time, it also places pressure on the ability of LPA planners to critically understand and evaluate what is presented to them.

As will become clear, the ALGE and the Chartered Institute of Ecology and Environmental Management (CIEEM) are both working hard to improve the quality of data collected (CIEEM 2019a) and received by local ALGE members in local planning authorities. As a first step towards this, they jointly produced an Ecological Impact Assessment checklist (CIEEM 2019a), based on the need for data adequacy set out in clauses 6.2 and 8.1 of the British Biodiversity Standard BS 42020 (BSI 2013).

Table 1.1 The hierarchy of protected designated sites in the UK (RTPI 2019).

Site designation	Details
Sites of international importance	
Ramsar sites	Listed under the Convention of Wetlands of International Importance 1971 (as amended)
Special Protection Areas	Classified under the Directive on the Conservation of Wild Birds 1979
Special Areas of Conservation	Designated under the EC Directive on the Conservation of Natural Habitats and of Wild Fauna and Flora 1992 (the Habitats Directive)
Biosphere reserves	Designated by UNESCO
Sites of national importance	
National Nature Reserves, National Scenic Areas and Marine Nature Reserves	Declared under section 19 of the National Parks and Access to the Countryside Act 1949 or section 35 of the Wildlife and Countryside Act 1981 (England, Scotland and Wales); articles 16, 18 and 20 respectively of the Nature Conservation and Amenity Lands Order 1985 (NI)
Sites of Special Scientific Interest and Areas of Special Scientific Interest	Notified under section 28 of the Wildlife and Countryside Act 1981 (England); article 24 of the Nature Conservation and Amenity Lands Order 1985 (NI)
Sites of regional/local importance	
Local nature reserves and wildlife refuges	Designated by local authorities under section 21 of the National Parks and Access to the Countryside Act 1949 (England), article 16 of the Wildlife Order 1985 (NI) and article 22 of the Nature Conservation and Amenity Lands Order 1985 (NI)
Non-statutory nature reserves	Established and managed by a variety of public and private bodes (e.g., county councils, wildlife trusts, RSPB)
Sites of Nature Conservation Interest and County Wildlife Sites	These are non-statutory sites of at least county importance for wildlife which meet agreed selection criteria. The status of this type of site varies considerably

EC = European Commission; UNESCO = United Nations Educational, Scientific and Cultural Organization; NI = Northern Ireland; RSPB = Royal Society for the Protection of Birds

Table 1.2 shows the range of questions asked so that an LPA might tick off issues in any application. This has clear merit. However, it depends on the ability of the LPA to understand the quality and forms of replies provided by applicants. This, as will be seen later, is one of the Achilles' heels of the LPA sector. The very fact that there are significant shortfalls in both the capacity and competence of LPA planners (see Section 1.3.1 and ALGE 2013; Snell and Oxford 2022) means that the checklist raises as many questions as it answers. In addition, as individual LPAs all operate their own versions of validation checklists (Abrahams 2019) – the things that they

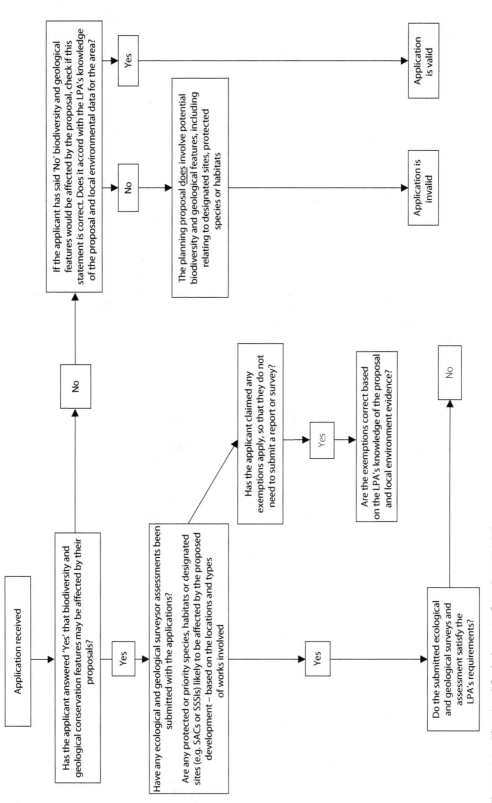

Figure 1.1 The simplified planning flowchart (ALGE).

10 | PROTECTED SPECIES AND BIODIVERSITY

want to see in an application – and not all appear to match Table 1.2, there is an ongoing problem. Until LPAs are able to fully understand what is submitted to them in planning applications, it means that there is little chance of a truly objective, critical appraisal of the sort set out by ALGE and CIEEM. Understanding what LPAs say they want is an important first step.

Table 1.2 The CIEEM/ALGE (CIEEM 2019) checklist for ecological data accompanying planning applications (ALGE and CIEEM).

	EcIA Criteria (to ensure decisions are based on adequate information in accordance with Clauses 6.2 and 8.1 of BS42020:2013)	Yes No n/a	Paragraph reference number(s)
Pre-app/scope	1. Where pre-application advice has been received from the LPA and/or an NGO and/or statutory body (e.g. NE DAS, NRW DAS), it has been fully accounted for in the EcIA		
	2. The scope, structure and content of the EcIA is in accordance with published good practice		
Surveys, Sites, Species and Habitats	3. Adequate and up-to-date: a. Desk study has been undertaken b. Phase 1 habitat survey (or equivalent) has been undertaken c. Phase 2 ecology surveys have been undertaken (where necessary)		
	4. All statutory and non-statutory sites likely to be significantly affected are clearly and correctly identified		
	5. All protected or priority species and priority habitats likely to be significantly affected are clearly and correctly identified, and adequate surveys have been undertaken to inform the baseline		
	6. Any invasive non-native plant species present are clearly and correctly identified		
	7. Where a separate PEA Report states that Phase 2 ecology surveys are required, these have been undertaken in full and results submitted with the application (or lack of such surveys is justified)		
Impacts and Effects	8. The assessment is based on clearly defined development proposals along with relevant drawings/plans (and any plans used are the same version number as those submitted with the application) or 9. The residual ecological effects are considered to be not significant at any geographical scale irrespective of the detailed development proposals, and the assessment is based on a worst-case-scenario		
	10. The report describes and assesses all likely significant ecological effects (including cumulative effects) clearly stating the geographical scale of significance (where relevant)		

	EcIA Criteria (to ensure decisions are based on adequate information in accordance with Clauses 6.2 and 8.1 of BS42020:2013)	Yes No n/a	Paragraph reference number(s)
Mitigation, Compensation and Enhancement	11. The mitigation hierarchy has been clearly followed		
	12. The report: a. Clearly identifies the proposed mitigation and compensation measures, and explains how these will adequately address all likely significant adverse effects b. Includes, where necessary, proposals for post-construction monitoring c. Recommends how proposed measures may be secured through planning conditions/obligations and/ or necessary licences		
	13. A summary table of proposed mitigation and compensation measures has been provided		
	14. The need for any mitigation licences required in relation to protected species is clearly identified		
	15. Proposals to deliver ecological enhancement/Biodiversity Net Gain have been provided		
Competence/Good Practice	16. Limitations of the ecological work have been correctly identified and the implications explained		
	17. All relevant key timing issues (e.g. site vegetation clearance or roof removal) that may constrain or adversely affect the proposed timing of development have been identified		
	18. All ecological work and surveys accord with published good practice methods and guidelines OR deviation from such guidelines is made clear and fully justified, and the implications for subsequent conclusions and recommendations made explicit in the report		
	19. All ecologists and surveyors hold appropriate species licences (where relevant) and/or have all necessary competencies to carry out the work undertaken		
Conclusions	20. The report clearly identifies where the proposed development complies with relevant legislation and policy, highlighting any possible non-compliance issues, and highlighting circumstances where a conclusion cannot be drawn as it requires an assessment of non-ecological issues (such as socioeconomic ones)		
	21. The report provides a clear summary of losses and gains for biodiversity, and a justified conclusion of an overall net gain for biodiversity		
	22. Justifiable conclusions based on sound professional judgement have been drawn as to the significance of effects on any designated site, protected or priority habitat/species or other ecological feature, and a justified scale of significance has been stated		

1.2 What do LPAs say they want?

Most LPAs – as would be expected under their biodiversity duties – have well-developed biodiversity policies and strategies, which include the relevant sections of legislation. All note that these duties need to be applied in submitting and reviewing planning applications.

For anyone wishing to submit a document for planning purposes, it makes sense for the LPA to provide simple, clear guidance on what it expects. However, this varies widely between LPAs (Abrahams 2019). For example, Huntingdonshire District Council, a small district Planning Authority, identifies its biodiversity duties in policy LP 30 as part of its local plan up to 2036 (Huntingdonshire District Council 2019). Under LP 30 it refers to some of the procedural steps that any application must follow. Some of these are set out in a short checklist. And in practice, these steps would seem to be discretionary rather than mandatory: observed more in principle than practice.

By contrast, the sorts of clear checklists and guidance provided for applicants by Chichester District Council are very helpful (Table 1.3). Chichester's concise 4-page resume sets out what it wants, why it is wanted and how these requirements should be met in any submission. It also says why it would fail an application. Details, and a clear checklist follow in 5 appendices. A similar approach is followed by Chelmsford District Council. Chelmsford's appendices provide more detailed guidance on survey times and outline methods. Both Chichester and Chelmsford refer applicants to the Governmental Natural England and DEFRA species survey guidance for detailed methodologies. In contrast, Scarborough District Council (since subsumed into North Yorkshire Council) has a dated Biodiversity Action Plan (2004) and provides little by way of biodiversity guidance; it refers applicants to the Wildlife Assessment Check (WAC) for survey details (www.biodiversityplanning.org).

As Table 1.4 shows and Abrahams (2019) noted, there is little consistency in the approaches of councils, risking variation in the quality and robustness of the data sought, provided or received. That in turn may impact on how well what comes across the desk is understood.

In order to try to help planners, especially those with limited staffing levels, such as Scarborough/North Yorkshire, the Biodiversity in Planning's WAC (PBP 1, see section 2.3) is being promoted as a helpful tool to guide applicants along the decision stages in order to end up with a valid application. Ideally, it will help preclude errors and insufficiencies in an application, resulting in an easier job for planners. This is examined in Chapter 2.

1.2.1 Why reliable data matters in the planning system

As with many elements of public discourse, where there are elements of law and contention, the planning system is essentially an adversarial one. This places the quality and reliability of data at the very heart of the matter.

At its simplest, an individual, group or company may wish to develop or alter part or all of a site. This could be a building, a vacant plot or a larger area of land. Details of

Table 1.3 Chichester protected species check list.

Proposals for development that will trigger a protected species survey/priority species survey and assessment	Please tick Yes / No	Species likely to be affected and for which a survey will be required
		Bats / Barn Owl / Breeding birds / GC Newts / Otters / Dormouse / Water Vole / Badger / Reptiles / Amphibians / Plants / Wintering & migrating birds / Invertebrates / Other BAP species
Demolition or conversion of agricultural or derelict buildings in a rural setting		Bats •, Barn Owl •, Breeding birds •
Development involving field hedgerow or woodland removal		Bats •, Breeding birds •, Dormouse •, Badger •, Reptiles •, Plants •, Invertebrates •
Proposals affecting large old trees		Bats •, Breeding birds •
Works within 50 m of a pond		Bats •, GC Newts •, Amphibians •, Other BAP species •
Works on or immediately adjacent to watercourses including rivers, ditches and rifes		Bats •, Otters •, Water Vole •, Amphibians •, Plants •, Invertebrates •
Proposals affecting mature or overgrown gardens, rough grassland, derelict land allotments of an area over 0.29 ha		Bats •, Breeding birds •, GC Newts •, Badger •, Reptiles •, Amphibians •
Proposals affecting flower-rich meadows or parkland		Bats •, Breeding birds •, GC Newts •, Badger •, Reptiles •, Plants •
Proposals involving flood lighting within 50 m of woodland, water, hedgerows or obvious connecting lines of trees		Bats •, Barn Owl •, Breeding birds •
Proposals affecting or adjacent to heathland		Breeding birds •, Dormouse •, Reptiles •
Proposal site within proximity of a known location of protected species		As records indicate, please consult Sussex Biodiversity Records Centre

Table 1.4 The differing expectations and requirements of 4 LPAs

	Planning Authority				
	Huntingdonshire District Council (HDC)	Chichester District Council (CC)	Chelmsford District Council (CDC)	Scarborough Borough Council (SBC)	Comments
Cover sheet	*Non-detailed	*	*	*	
Biodiversity checklist	*	*	*Detailed	–	
PEA recommended	*	*Required	*Not named but details match	*Required	
Use local record centre	*Not mandatory	*Required	–Not mandatory	*	
Use impact risk zones	*	*	–	–	
Use qualified surveyors	*Required	*	*Required	*Required	
Survey at correct time of year	*Details given	*	*Details and references given	*Required; cites WAC	Some provide text, others a chart. SBC refers to WAC
References for methods provided	–	*Government website	*Government website	*Cites WAC	No consistent references or levels of detail
Demonstrable net gain expected	*Required	*Required	*Required	*Required	Older (CC) cites enhancement; more recent state net gain
Follow county/LPA biodiversity plans	*Fit in with county BAP	*	*Fit in with plans and county plans	*Assumed	SBC has own (2004) BAP

(source: LPA websites). *needed;   – not mentioned. PEA = Preliminary Ecological Assessment; WAC = Wildlife Assessment Check; BAP = Biodiversity Action Plan.

that proposed change form the core of a planning application submitted to the LPA. The application will normally be accompanied by a range of subsidiary documents which set out in detail how the application appears to meet local planning legislation and policies.

Once submitted, and then validated, the application documents are open to public scrutiny. This is also when the adversarial element can kick in, as the data presented, and how they are interpreted, can be called into question by both the LPA and those with an interest in the outcome. These may be the immediate neighbours, or a range of groups opposing larger and more contentious applications. It is at this point that the reliability of the data submitted to the LPA may be queried. Being able to understand the implications of possible biodiversity changes caused by a planning application sits squarely in the job description of local authority planners (RTPI 2019). This may be a problem as LPA staff levels continue to shrink (Oxford 2012; ALGE 2013; HoL 2018; Snell and Oxford 2022).

It is a common political axiom: there is 'wastage' to trim in any bureaucracy. In the UK, central and devolved governments have cut many posts across all sectors at national and local levels, including planning. Among those sections of the LPA 'fat' most targeted have been the planners, the source of 'burdensome' delays to developers (Gove 2018). The ALGE has watched, and reported (ALGE 2013; HoL 2018; Snell and Oxford 2022), as specialist ecologists have been lost from LPAs. With this has come a reduction in the capacity of the authorities to effectively evaluate planning applications from an ecological and biodiversity standpoint. LPA ecologists have not been immune from the metaphorical axe, with more work expected to be done by those that remain.

Whether or not there is any 'fat' or slack in the planning divisions of most LPAs is debateable (Oxford 2012; ALGE 2013; HoL 2018; Snell and Oxford 2022). Regardless of size, the UK planning system operates on a simple premise: evaluating whether a planning application may or may not have any negative impacts (on people, on landscapes or the setting of buildings, as well as on protected species and general bio or geodiversity) before agreeing or refusing permission. Understanding what these impacts might be, and if these will be significant, minor, absent or positive, requires three simple elements in order to assess each application.

- The first is the provision of good, reliable and valid datasets to planners. Without these, the merits of the case being put forward by the developer cannot be assessed.
- The second is the availability of detailed and suitable guidance for use by planners that will allow meaningful evaluation of those data and their claims.
- The third is a large enough body of suitably qualified LPA planners or available staff, with enough time, to use that guidance to make informed decisions and provide viable and credible recommendations to councillors who will normally ultimately decide on an application.

If those three elements are met, then planning assessments might be expected to contribute to an authority's and government's policy of no net loss or biodiversity net gain. If one or more of these is not met, then the whole pack of cards falls. Clearly, such a failure would be a serious issue.

Of the three elements, the lack of LPA ecologists is already a problem. The House of Lords (HoL 2018) reported that the number of LPA ecologists had fallen by 60%, leaving many LPAs without adequate internal cover. Leaving all biodiversity evaluations to the 10% of planners with ecological qualifications (ALGE 2013; Snell and Oxford 2022) is not a realistic answer, as they too are stretched thinly.

If there are few suitably experienced or qualified staff in place, then it is critical that the other two elements are met: good guidance is available to evaluate the biodiversity aspects of planning submissions and the provision of suitable data by developers. Without both, there is no way that applications can be determined effectively. ALGE (2013) and Snell and Oxford (2022) stated that biodiversity data in planning submissions are rarely as robust as claimed. Without good guidance that tells planners what to look for in data accompanying a planning application, then the field is wide open to mediocre submissions and poor-quality data, and therefore the risk of ill-informed decision-making. For biodiversity, protected species and ecosystem services, that can only spell trouble.

1.3 ALGE, ecology and planning

All aspects of planning legislation delivered at the LPA level need specialist input. Biodiversity, ecology and protected species are no different. In order to deliver statutory biodiversity obligations, such as the NERC Act 2006 (HMG 2006) and NPPF (HMG 2021), LPAs need a sound ecological skill base to confirm the provision of credible data both for LPA use and when summary data are used for government policy purposes.

This means that planning authorities might be expected to have access to specialist, ecologically trained staff, much as there are officers specialising in listed buildings, noise, archaeology, public rights of way and other aspects covered by planning policy. Historically, this would have meant having either staff ecologists or access to external resources through a retained contract (ALGE 2013; HoL 2018; ENDS 2019; Snell and Oxford). However, it is clear that, rather than expanding numbers over time in line with government's legislative requirements and biodiversity commitments, the numbers of ecological specialists has dropped (Oxford 2010, ALGE 2013, 2016). ENDS (2019) suggested a slight rise in numbers in its subsample of planning authorities replying to a freedom of information requests. That would not begin to offset the documented reductions. Snell and Oxford (2022) confirmed the paucity of LPA biodiversity specialists.

Perhaps for those wielding cuts, the reduction has been because it is an easy topic area: everyone knows about wildlife, so a bit of knowledge, allied with common sense, should suffice for planners to make sensible decisions? And, as it might be reasonably assumed, if only a tiny fraction of planning cases would have any possible impact on plants and animals, then for these odd cases a bit of bought-in skills would be enough?

In spite of such optimistic, and baseless, assumptions, the scale of the task is not small. As LPA planning decisions have the potential to be challenged on factual and procedural bases, and may possibly end up at public inquiry, ecological and other decisions need to be well founded. As a result, LPAs employ a wide range of specialist staff, including ecologists, who mainly operate under the aegis of the planning department.

In its 2013 review of the ecological capacity and competence available in English LPAs (ALGE 2013), ALGE looked at the scale of input required every year. In the year ending June 2013 (although a decade ago, it is still the year with the most recent detailed data), district level LPAs determined roughly 418,000 planning applications. Tyldesley and Bradford's (2012) study of the approximately 14,000 major planning applications submitted to English LPAs estimated that biodiversity was a material consideration in 29% of them. They suggested that at least 4,000 major applications each year were likely to require substantial scrutiny and input on biodiversity issues. That required skills, knowledge, time and good guidance documents to sift the wheat from the chaff. All of which seem to be in short supply (ALGE 2013; Snell and Oxford 2022).

While major applications tend to hog the limelight, ALGE (2013) noted that many of the thousands of smaller-scale applications were also likely to have significant implications for protected species and for priority habitats and species. For instance, if only 5% of the total non-major applications (e.g., roughly 20,000 of 418,000) were likely to affect important habitats and species, this would mean that there were over 20,000 applications each year where input from a competent ecologist would lead to a better outcome for biodiversity conservation (ALGE 2013).

Clearly, that is not a small figure, and even though these are old data, by implication it places the statutory authority in a bit of a spot if staffing is not adequate to meet demands, and especially so if capacity and competence are at issue (ENDS 2019; Snell and Oxford 2022). This would affect both the ability to evaluate an application and to understand any guidance that might be available to help do this, especially if that guidance were itself limited or made assumptions about those likely to use it.

The ALGE (2013) review painted a less than rosy picture, one that was not better in 2016 (ALGE 2016; HoL 2018) or 2019 (ENDS), but as it is the most detailed available, it is a useful to set the scene for the problems that faced LPAs then and, by extension, now (Snell and Oxford 2022).

In order to understand the breadth and quality of the ecological resources and competence available within the LPA sector, ALGE undertook two surveys of its LPA members. It published its results in November 2013. Although updated in part, it remains the most comprehensive review of the resources and problems that face LPAs and the establishment of a credible baseline for biodiversity data and tools, such as net gain, which are predicated on a better resource than is actually available at the LPA level. The most recent ALGE survey (Snell and Oxford 2022) shows no improvement in their ability to deliver biodiversity net gain.

One of the first elements considered in the 2013 ALGE study was whether having good data mattered. If not, then worrying about resources and capabilities might be a cosmetic concern but not an actual issue.

ALGE referred to a DEFRA-sponsored study by Tyldesley and Bradford (2012). Looking at the quality of outcomes for biodiversity in planning when expert ecological advice was available at LPAs along with sufficient ecological data (and expertise allows differentiation between good and bad submissions), Tyldesley and Bradford (2012) found that both were needed. Poor skills and knowledge, coupled with poor submissions, was a negative mix, as possible impacts and data problems were missed without in-house ecologists. Similarly, positive opportunities were also likely to be missed. DEFRA (2013) also recognised that problems with the capacity and abilities in the planning system slowed decision-making and often underplayed the effects of an application on biodiversity. These findings influenced ALGE's (2013) study.

ALGE (2013) considered what would be a positive mix for dealing with biodiversity in planning applications. This it called *technical resilience*, and was a combination of:

a) an appropriate level of *professional competence*: the ability to undertake technical ecological assessments and to make informed, and sound, recommendations;
b) enough *capacity* to deliver the workload of complying with statutory obligations and policy workloads.

Having enough capacity, but no competence, would be as unsuitable as having competence but no capacity. Similarly, having only some of both would also fall short. Perhaps the most important aspect was technical competence; without it, no matter how many staff, the outcome would be irrelevant and fall short of statutory duties.

In order to understand what level of expertise might be needed, and how planners viewed their own skills, ALGE used the CIEEM Competency Framework (CIEEM 2019) to identify the necessary level of competence that would be expected to effectively assess the range of the biodiversity-related material that came across their desk. These would range from:

- basic (simple understanding of concepts and ideas and able to do basic assessments under guidance), to
- capable (enough ability/experience to do standard assessments but would still defer to others for complex cases), to
- accomplished (enough knowledge and experience to do non-standard work and consider alternative options) through to
- authoritative (peer-recognised for knowledge and skills).

For the 88 English LPA ecologists that responded in the 2013 study, the picture was not impressive. They were asked what level they would expect planners should be able to work at in 4 main categories:

1. ecological surveys;
2. scientific method;

3. environmental management;
4. impact assessment, habitat regulation assessments and environmental legislation and policy.

The response was that planners should rank as capable or accomplished for these categories and their subdivisions. When asked their own levels, almost all felt their competence was lower: basic. Only when thinking about legislation and policy did planners rate their own skills as capable.

Being able to determine the appropriateness of methods and surveys used (categories 1 and 2), whether species or habitats were identified correctly, whether the field data were adequate or interpreted correctly or decide if conclusions are warranted, all require more than the minimal (basic) technical levels reported by almost all planners. Instead, there was heavy reliance on ecologists or bought-in resources, if available.

There was the same expectation for environmental management (category 3) – that planners should have capable or accomplished skill levels, but this was not reflected in their own self-assessments. A total of 90% thought their knowledge level was basic. Again, this was backfilled by ecologists, regardless of duration of experience of a planner.

The same applied to areas of legal coverage of ecological and biodiversity issues (category 4).

As ALGE (2013 S1.15 and 4.2.5) put it, 'there is an ecological skills gap in the planning system'. These gaps were most keenly felt in:

- review of ecological survey methods;
- scientific analysis and interpretation of ecological data and information;
- environmental management (measures for mitigation, compensation and enhancement);
- ecological impact assessment;
- habitat regulations assessment;
- compliance with and enforcement of environmental legislation and policy;
- understanding and application of the mitigation hierarchy.

These are pretty fundamental gaps, given that LPAs have statutory obligations to:

- have due regard to biodiversity conservation in the exercise of the planning functions;
- consider priority habitats and species in their decision-making;
- consider and consult on the implications of development affecting Sites of Special Scientific Interest;
- assess the effects of development on sites and species protected by national and European law;
- consider all relevant environmental information (for Environmental Impact Assessment developments) prior to the grant of planning permission.

Then there are problems that affect the very basis on which biodiversity, ecosystem services and, from 2023/2024, biodiversity net gain would be evaluated (Rampling et al. 2023). ALGE (2013) suggested that

> in the absence of appropriate specialist assistance, many planning authorities will be unable to competently and effectively undertake the technical assessment of ecological issues set out in Clause 8.1 of BS 42020. As a result, it is highly questionable whether a planning authority could confidently conclude that they have effectively (and lawfully) addressed the statutory obligations.

That is not a criticism of LPA ecologists, but rather a statement that no matter how well intentioned they are, having too few in place compromises the volume and quality of what can be provided. DEFRA (2013) put it in another way:

The lack of ecological awareness amongst most planning officers who find it difficult to recognise and place appropriate weight on biodiversity issues in the balance of planning judgment, especially without expert advice; these impediments can then lead to:

a. failure to recognise biodiversity issues as material during the consideration of planning applications.

b. decision-makers giving less weight to biodiversity issues in decision-making than attached routinely to more frequently encountered planning issues such as economic regeneration, affordable housing, design, layout and traffic considerations.

Clearly, planners needed support from some sort of ecological reference: human or text. ALGE (2013) noted several kinds of support. The simplest remedy would appear to be the in-house ecologist. But roughly 65% of LPAs did not employ ecologists. Instead, many LPAs appeared to pool resources to gain access to ecological staff employed by another LPA. Nonetheless, a third had no access to any internal or external advice. That would not be a sound basis for informed decision-making by planners with just basic skills in this important area. The position had not improved by 2021 (Snell and Oxford 2022).

Given the self-reported limitations of ecological competence in planning staff, it would be expected that planners would turn to other sources to fill these gaps. Yet, when judging if biodiversity was likely to be affected in an application, a surprising 75% of planners said that they trusted their own professional judgement in spite of the admission by the vast majority that this was basic – the lowest competence category. Quite why they felt so confident, when other sources of guidance were also not used, is unclear. Roughly 30% used Natural England's ecological standing advice for planners, at the same time as 47% stated they did not. If not relying on Natural England's standing advice, then they clearly did not get advice from Wildlife Trusts, as only 8% rated a Trust as their first port of call in lieu of in-house support, through some sort of service-level agreement. Even fewer bought in skills from consultancies. On that basis, planners were making decisions for which they were unable, or did not have competency, to determine and validate their decisions. That remained true in 2021 (Snell and Oxford 2022).

This very mixed picture indicated that although in-house ecological staff were the preferred choice for advice, the basis on which planners made their decisions was far from certain, and the results were in line with DEFRA's (2013) negative assessment. That quality and capacity gap remains (Robertson 2021; Rampling et al. 2023).

If ALGE members (professional ecologists within the LPA structure) are so important, it seems odd that LPAs might see them as an 'optional extra' (ALGE 2013). Oxford (2012) reported that ecologists suffered severely in LPA cuts in 2011, with 5–100% of budgets lost. A total of 26% of LPAs lost 1 ecologist, and 18% lost 2–4 ecological staff. By 2013 the process was continuing apace and had not improved significantly by 2018, or later (ENDS 2019; ALGE 2020; Snell and Oxford 2022).

In giving evidence to the House of Lords on the NERC Act (HoL 2018) and the capacity of Natural England to provide advisory support roles to LPAs, ALGE noted that 90% of LPA planners lacked ecological qualifications, had little relevant training and reaffirmed their incapacity to address issues in more than the most basic of ways. This was not being filled by Natural England staff, or its antiquated guidance (ALGE 2016a). In its evidence to the same committee, CIEEM (2018) reaffirmed this gap in competence and capacity in LPAs, noting that:

> planners are ill-qualified to make biodiversity decisions and are not competent to do so; they do not claim to be so either, but the requirement falls to them due to lack of resources. (CIEEM 2018)

CIEEM suggested that LPAs should be required to employ, or contract, a competent ecologist to advise them on all aspects of policy or duties that could impact on the natural environment. Neither ALGE nor CIEEM were supportive of the abilities of LPAs to deliver their statutory roles.

By 2018, ENDS (2019) reported that only 25% of councils in England employed ecologists, down from 35% in 2013, and for some this was a highly thinned resource in absolute numbers. Just 10 authorities employed 28% of all LPA ecologists. Some LPAs, such as Greater Manchester's local authorities, supported a joint ecological unit, while others, such as in Surrey, were stretched to the extreme. Here, one full-time ecologist supported 11 district and borough councils.

Considering the very low level of self-assessed ecological competence in planners, it might be expected that ecological staff would be relatively highly qualified. In the UK, membership of the CIEEM is seen by many as a threshold of ecological competence, and reaching chartered ecologist status represents one of the two pinnacles in CIEEM, equivalent to the level of accomplished. ENDS suggested that 49 (34%) ecologists were chartered ecologists. According to CIEEM, having suitable in-house expertise was critical as:

> without it we think that they are vulnerable to challenge – in strategic plan making and development decision-making. That can make planning

authorities risk averse, but also non-compliant with legislation and policy. (CIEEM in ENDS 2019)

The arrival of net gain on the planning desk promised to make the planning job no easier and to place further emphasis on the availability of suitable competence and capacity. As the head of land use planning at the Wildlife Trusts put it:

> ensuring that authorities have access to the right level of ecological expertise will be key to the policy's success. Net gain will mean that local authorities need to understand the information provided by applicants and their consultants and interpret complex net gain calculations to assess whether they will lead to a real long term net biodiversity gain for nature.
>
> Local authorities will also have a role in monitoring and enforcement of net gain delivery. Without adequate resources ... net gain risks failing nature. (Young in ENDS 2019)

The housebuilder Redrow, agreed:

> Net gain will add to existing local authority ecologists' workloads as they will need access to suitably qualified and experienced ecologists to review biodiversity net gain information submitted by developers. (ENDS 2019)

1.3.1 Capacity, competence and LPAs

Given the relative paucity of ecologists across LPAs, imposing a rigorous demand to operate net gain calculations by early 2024, as part of Government's 25 Year Plan and 2021 Environment Bill, posed a challenge. Lacking competence and capacity to perform existing roles, adding new loads to LPAs threatens data quality and the risk of letting through errors, rather than providing tight control over what is allowed to pass (Robertson 2021; Snell and Oxford 2022; Rampling et al. 2023). Add to this the problems of data quality received by LPAs, and there is the possibility of a perfect storm.

Throughout the previous 10 years, ALGE (2020) had been concerned about the quantity and quality of the ecological resource base available to LPAs. The arrival of the 25 Year Plan in 2018, and the 2021 Environment Bill, presented fundamental challenges to LPAs and to ALGE.

Early in 2020, ALGE produced a report looking at the Plan and the Bill. Having previously said that LPAs were incapable of delivering on their existing statutory duties, ALGE (2020) was not inclined to alter its stance. While welcoming the recognition from central government that local government was key to delivering its environmental ambitions at the local level, ALGE (2020) stated that there was a mismatch between aspiration and reality. As it put it:

> Currently ALGE feel there is a gap between the ambitions of the 25 Year Plan and the local delivery mechanism, skills and capacity available to achieve the Plan's outcomes. Many Local authorities have neither the capacity nor the capability to deliver the 25 Year Plan

Imposing a mandatory requirement onto a body lacking capacity or competence was not an obvious step. ALGE restated its concerns: that planners were not competent enough to assess applications involving net gain, nor to consistently apply the principles underpinning the concepts. Nor was ALGE convinced that the ecologists in place within LPAs would have either the right skills or capacity to deliver the new specialist input required for aspects of the 25 Year Plan and the Environment Bill. Echoing its 2013 position, it concluded that:

> It is ALGE's view that the 25 Year Plan and Environment Bill, whilst laudable strategic policy documents, lack a delivery mechanism for local authorities, many of whom lack the capacity or capability to implement key areas they are responsible for. (ALGE 2020)

In 2021, DEFRA funded a further review of ALGE's capacity to deliver biodiversity net gain (Snell and Oxford 2022) in England. The review echoed almost all of the 2013 findings, and reiterated the 2020 conclusions. These included the following points.

- Capacity and competence in English LPAs were inadequate to deal with existing biodiversity-related workloads, let alone an added biodiversity net gain burden.
- Only 5% of respondents said their current ecological resources were adequate to deal with applications concerning biodiversity.
- Less than 10% thought their current competence and capacity were adequate for biodiversity net gain delivery.
- Approximately 90% of LPA planners lacked ecological qualifications and had received little relevant training.
- Approximately 26% of LPA planners have no access to ecological advice.
- Even though unqualified, 73% of planners use their own judgement in assessing applications.
- More than half used DEFRA/Natural England standing advice guidance in helping to evaluate applications.
- Most applications lack adequate biodiversity details that are needed to evaluate planning applications.

If LPAs cannot be relied upon to deliver basic planning assessments, let alone biodiversity net gain (ALGE 2020; Snell and Oxford 2022), then does that also apply to the material presented to them too?

The mix of limited competence in planners, a shortfall of ecologists in-house, a rising demand for inputs to planning applications and the burgeoning multiple requirements of the 25 Year Plan and the Environment Bill and biodiversity net gain all threaten to make planning evaluation and biodiversity net gain assessment (Rampling et al. 2023) an ever-riskier place.

If LPA planners are poorly equipped to deal with many of the detailed biodiversity aspects of planning applications, would planners be able to recognise poor planning applications and just how common these were? As importantly, is there any sort of consistency in the material required by planners, and are there reference standards that they use or recommend that are suitable for need, by them or by applicants? This can be addressed in two ways:

- through the material set out in Validation Checklists; and
- through the reference sources used and recommended in those lists and ancillary LPA guidance.

1.3.1.1 Lists and checklists

Before formally arriving on the LPA desk, most planning applications will normally have involved a limited amount of pre-discussion between the would-be developer and the LPA (BSI 2013; CIEEM 2019a), as potential applicants sound out LPA staff on their expectations and possible concerns, and the sorts of information that the LPA might reasonably want to receive as the competent authority in the planning process. This is a standard process for those LPAs listed in Table 1.4 (page 14) for example.

Pre-discussions would include reference to LPA policies and statutory obligations and to formats and contents of guidance documents, such as the Ecological Impact Assessment Standard of CIEEM (2019a) or local validation checklists (Abrahams 2019), which set out the list of things to be covered in a planning application.

It might be expected that validation checklists would be very similar for all LPAs, but this is far from the case. Under the NPPF (HMG 2021), LPAs are required to take a proportionate approach to data requirements in submissions (basically that old issue of burdens) and to provide guidance to planning applicants. How to do this was not strictly defined.

Without DEFRA providing unequivocal guidance, LPAs developed their own versions of validation checklists, mainly based on the draft ALGE (2007) guidelines for planners. Never completed, the ALGE template has been used and adapted by a range of LPAs to guide their information requirements. And with it has come a whole range of local variants of very differing quality (Abrahams 2019).

The second touchstone came with the development of the national level 2014 DEFRA/Natural England standing advice on the DEFRA/Natural England website, which set out some of the requirements that a planning application might be expected to contain. As LPAs began to adopt these, usually on top of ALGE templates, the result was the beginning of a broth of validation checklists, each flavoured by the inclination or experience of its many authors. Top this up with the NPPF requirement that LPAs review, update and publish their validation checklist requirements every two years, and you have the starter kit for mayhem.

For the consultant, often working across several LPA boundaries, it became clear that what was acceptable or expected in one LPA area was not acceptable in another. Abrahams (2019) observed that even a European protected species, such as the Great Crested Newt *Triturus cristatus*, received very different validation checklist guidance between LPAs.

The result, as Abrahams (2019) noted, was confusion. There is the real risk that LPAs might promote – no matter how well intended – negative impacts for protected species and loss, rather than gain, in biodiversity features as a result of an inadequate planning process and data submissions.

By the end of those pre-application discussions with their LPA, the applicant will hopefully be better informed, and the LPA will have come to a clearer understanding of the potential, and possible, issues that they will need to check when the application finally arrives.

The submitted application letter normally includes a range of supporting documents. The first, the formal application document, will usually include a statement of whether or not there will be a biodiversity or ecological impact. The larger the proposed development, the more likely that this section will be filled in as a possible effect on biodiversity. Often, the 'no' entry in this section will be contradicted by other documents submitted with the application (Treweek and Thompson 1997; Drayton and Thompson 2013).

For all but the smallest domestic house extension (and even this may be affected by a bat roost; Chichester District Council (2018)), there is a risk that there may be a biodiversity impact, including on protected species, directly or indirectly, or as part of some wider cumulative impact (PBP 2021; CIEEM 2018). This places great emphasis on soundly prepared desk and field data accompanying an application. Without this, there is no basis for confirming or denying a potential impact, or whether or not the application is in compliance with local and national policies and legislation.

2. Guidance and Interpretation

If one of the goals of a planning application is to determine the possibility, or otherwise, of a potential impact associated with that application, then it is critical that the methods used are appropriate and the results are unequivocal. For biodiversity, protected species, ecosystem services or natural capital, this relies on the availability, and use, of suitable guidance by both applicants and planners as well as the capacity and capability of LPA staff to interpret this. As observed by ALGE (2013, 2020; Snell and Oxford 2022), this is not guaranteed.

In this chapter, I will look briefly at the range of guidance available to applicants and planners in England and consider how this might be used, as well as the caveats that should accompany it.

Biodiversity guidance for applicants and planners

There are three main sorts of guidance available to planners and applicants in England:

1. Department of Environment, Food and Rural Affairs (DEFRA) and Natural England (NE) guidance – NED for short- issued under the Natural England banner.
2. The Partnership for Biodiversity in Planning (PBP) Wildlife Assessment Check (WAC).
3. Individual species and group-specific best practice methods (SGN) from the PBP.

Both the DEFRA/NE (NED) and WAC guidance are meant to be distillations of planning practice and components of the best practice methodologies. For the potential user, the challenge is to understand the extent to which they provide the right level of detail and the caveats that should accompany their use.

Both NE and WAC guidance skim through significant aspects of detail, with the risk that they offer unwarranted reassurance, and with it the possibility of accepting inappropriate datasets that fail to raise warning flags and preclude their use for biodiversity net gain calculations. Both refer to some of the main best practice methods, but neither fully cover the breadth or nuances of individual methodologies. The SGN comes an important part of the way.

2.1 NE guidance 2014–21

Historically, many planning consultations would be viewed by conservation staff in each national statutory conservation agency in the role of a statutory consultee, usually at the level of county officers (Sheail 1998). This is no longer the case, and the range and number of consultations has diminished. In practical terms, in England this means that a reasoned evaluation of an application forwarded to NE specialist staff by an LPA has been replaced by a standard letter referring the local authority to the NED Guidance produced by NE, which now constitutes their opinion in all but the most major of cases.

2.1.1 What is the NE guidance, and how did it come about?

There are three strands of NE advice that were originally drafted and that were periodically revised between 2014 and 2021:

1. Protected Species and Development: advice for Local Authorities (PSD).
   (https://www.gov.uk/guidance/protected-species-how-to-review-planning-applications). This introduced LPAs to NE's expectations of their role in looking after protected species in the planning process. First produced in 2014, it was updated often until late 2021.
2. Preparing a Planning Proposal for protected species (PPP).
   (https://www.gov.uk/guidance/prepare-a-planning-proposal-to-avoid-harm-or-disturbance-to-protected-species). This was aimed at developers, helping them to avoid harm to protected species. Like the PSD, small changes were made most years prior to 2022.
3. Standing Species Advice (SA).
   (https://www.gov.uk/topic/planning-development/protected-sites-species). This was a more detailed set of guidance for individual protected species, supplementing some of the outline material in the PSD. Like the PSD and PPD, some of the SA was altered over the years.

NE's objective was to ensure that developers did the right thing: that suitable material would arrive on the planner's desks and that development could proceed without harm. If there was a risk, that would be clear to all. Whether this was achieved will be covered in outline in the following section. Refer to the Appendix for an in-depth review of the reality.

The PSD

In replacing consultation with NE staff with generic guidance, the pre-2022 PSD (Natural England 2014-2021) advice stated:

> Local planning authorities (LPAs) should use this guide to assess whether a planning application would harm or disturb a protected species. It will help you decide if you can give planning permission.

That presumed a set of skills and capacity that ALGE (ALGE 2013, 2020; Snell and Oxford 2022) has noted was largely missing in LPAs. The pre-2022 PSD replaced regular consultation, noting:

> This is Natural England's 'standing advice'. This is general advice that Natural England, as a statutory consultee, gives to LPAs. It:
> - avoids the need to consult on every planning application
> - helps you make planning decisions on development proposals.

To fill any void, the original 2014 PSD recommended: 'Use an expert, such as your local authority ecologist, to help you apply the standing advice to planning decisions if you're not a wildlife specialist'.

Perhaps because of the haemorrhaging of ecologists from local authorities, this was replaced in 2021 with the suggestion: 'You should get advice from a qualified ecologist to help you reach a decision if you need it'.

A later reference to expert advice has a hyperlink to NE's paid-for advice at £500 for a 90-minute session.

The only circumstances when LPAs might consult NE were given as follows:

> You must consult Natural England if a development proposal:
> - might affect a site of special scientific interest (SSSI)
> - needs an environmental impact assessment
> - needs an appropriate assessment under the Habitats Regulations.

In such circumstances, NE noted that:

> Natural England may:
> - object to a planning application if it's likely to harm a protected species on a SSSI
> - give you advice about a protected species affected by a planning proposal or on a specific issue that is not covered by this guidance.

To help LPAs, the PSD offered a short table (table 1 in the PSD advice) of species and species groups to look out for in a range of habitats. It was an odd list: short on marshland, moorland or mountain: all critically important habitats for protected species; for windfarms in particular, these were vital planning battle grounds. In table 2 of the PSD advice, NE provided a summary of suitable protected species survey dates expected in planning applications. Many of these differed from dates stated in the more detailed standing advice species guidance.

The overall objective of the PSD approach, and the summary tables, was to allow the LPA to determine if it had suitable data on its desk. As the PSD stated:

> If the information is not adequate you should ask for further information, such as further surveys or mitigation measures.

You can refuse planning permission if surveys:

- are carried out at the wrong time of year
- are not up to date
- do not follow standard survey guidelines without appropriate justification
- do not provide enough evidence to assess the likely negative effects on protected species.

The only potential way offered to determine this would be the NE protected species standing advice. NE stated:

If the proposal is likely to affect a protected species, planning permission may be granted where:

- a qualified ecologist has carried out an appropriate survey (where needed) at the correct time of year
- there is enough information to assess the impact on protected species
- all appropriate avoidance and mitigation measures have been incorporated into the development and appropriately secured
- a protected species licence is needed [and] it is likely to be granted by Natural England or DEFRA
- any compensation measures are acceptable and can be put in place
- monitoring and review plans are in place, where appropriate
- all wider planning considerations are met

The decision-making checklist (PDF, 162KB, four pages) can help support planning decisions.

The question that the LPA planner has to ask is: is the PSD and standing advice guidance enough to allow me to determine a decision on my own? This is especially true if the planner is low in competence and the LPA is also short on capacity.

PPP

Ideally, an applicant would have used the PPP advice to help structure their assessments, surveys and presentations, so that they would readily meet LPA requirements. That would include looking at the standing advice guidance. The objective of the PPP was to ease the route to development:

The LPA can ask you for the following.

A preliminary ecological appraisal to decide if you need to do a further survey where it's not clear:

- if there are species present
- which species they are
- if their numbers are significant to the species population as a whole.

More detailed surveys to provide an assessment of the potential effects of your development.

The LPA can request these because:

The LPA can refuse planning permission if the surveys:

- are carried out at the wrong time of the year, are not up to date or do not follow standard survey guidelines without appropriate justification;
- do not provide enough evidence for them to assess the likely impact on the species and the supporting habitat.

For an LPA planner, the common source is given as the Species Standing Guidance: the standing advice (SA).

Standing advice

The Natural England SA was intended to provide for the needs of planners and developers and was written as a 'how to' guide for individual protected species and groups. The SA framed its use in a series of paragraphs.

What it expected was introduced in the preamble to each SA document. The preamble for Badgers (Natural England 2015) *Meles meles* set out the SA stall for individual species in general terms:

Survey reports and mitigation plans are required for development projects that could affect protected species, as part of getting planning permission or a development licence. Surveys need to show whether protected species are present in the area or nearby, and how they use the site. Mitigation plans show how you'll reduce or compensate for any negative effects to protected species.

The advice went on thus:

This is Natural England's species standing advice for local planning authorities who need to assess planning applications that affect Badgers. This information can be used to decide what is needed for surveys and when assessing mitigation measures for Badgers.

That is a categorical statement about what is possible. Whether this is valid varies strongly between individual sets of SA. The SA continued:

Ecologists need to decide which survey and mitigation methods are right for the project they're working on. If this standing advice isn't followed, they'll have to include a statement with the planning application explaining why.

It is not just a simple matter of explaining why a choice was made, but also why it did not make any clear impact on the data and decisions reached from those data (BS 42020 (BSI 2013)). Limitations are a critical part of any analysis, and it should be clearly demonstrated why they mattered, or not. That is not a simple yes or no issue, as it has the potential to severely impact the baseline for subsequent monitoring and concepts such as net gain.

Taken at face value, NE was saying the SA is enough; little else was needed. That makes one or two assumptions: that the details are enough, that the LPA planners know enough, or that there is a source of ecological skills within the LPA that is on hand to plug those gaps. ALGE (2013, 2020) and Snell and Oxford (2022) have shown that the last two points are not valid. Whether or not the details were of themselves enough is open to question.

If a set of guidance has to be adhered to (see CIEEM 2021a), then it is important that it meets certain basic criteria that will allow anyone to evaluate the materials that are produced. The SA set of guidance used a series of high-level categories in almost all SAs:

- decide if you need to survey
- survey methods
- survey effort required
- assess the impacts
- mitigation and compensation methods
- mitigation measures.

For an LPA planner, additional – more detailed – headings specifying what is needed would be better; that way, planners can tell what was not done and perhaps why too. These would include:

- seasons/months to be worked, along with defined time(s) of day are specified, both to be worked and avoided;
- specification of suitable/unsuitable weather conditions, with limitations associated with unsuitable conditions noted;
- a clear statement of the minimum number of visits to be undertaken (when and how spaced over time) plus the maximum visits if this is a factor in interpreting data;
- number of workers needed if required;
- detailed methods and materials/equipment needed;
- distances to be sampled in/around potential sites;
- mapped data with walk routes, times and conditions stated;
- limitations specified, with why they matter and how these affect data collection or interpretation;
- the factual basis of mitigation options suggested.

With detail at that level – as expected by BS 42020 (2013) and CIEEM (2019a) – planners and their advisers would really be able see what took place and then would be better able to make a reasoned decision.

This suggests that much of the basic material risks being deficient. If so, where would an LPA planner go to check meaning or what was missing?

A long list of references, with commentaries on most, is available on CIEEM (2021a). Because these were missed in almost all SA, it was clear that the SA could not claim to be adequate or reliably self-contained. That was a basic problem for a planner. Even referring to that CIEEM list presupposed a degree of capacity or competence to follow through with it, which are both usually stated as absent within the LPA.

Was the individual SA guidance fit-for-purpose? In the Appendix, detailed reviews of SA guidance against specialist references – the sort of reference missing in total from the SA – showed that planners risked getting poor-quality data: the sort of data that rarely help to decide a case easily.

For example, the SA for the Badger *Meles meles* mandated neither distance, nor duration, nor frequency for any of the potential parameters. That made it hard to determine the claims of a developer to have followed the SA or ignored it for good reason. Many of the limitations that affect the potential methods, and their interpretation, were omitted from the SA guidance. Overall, this suggests that the pre-2022 Badger SA was not a robust set of guidance for use by planners as it omitted very basic elements that would help interpret a planning application.

What about other species? For reptiles (Natural England 2015a), rather like the Badger, the lack of depth in the SA precluded any meaningful use of the SA by LPAs when determining planning applications without recourse to additional sources of advice. If the LPA planner wanted to begin to understand the background to reptile surveys, and to try to get to grips with impacts and data reliability, then they had to look away from the SA. The paucity of suitable detail meant that the SA failed to meet its own claims.

For bats (Natural England 2015, 2020), it had been expected that the 2020 SA update would have produced a much clearer, simpler, better expressed set of guidance usable by LPAs and practitioners alike. The SA did not meet its own claims. It could not be used to evaluate an applicant's submission on its own. Both applicants, and more importantly, the LPA would need to refer repeatedly to Collins (2016) to make sense of needs and claims and their verifiability. It is unfortunate that the SA used terms in a way that Collins did not and that details differed between Collins and the SA. The limited references to bats and micro or small turbines failed to meet the needs of either side. To be useful, a separate section should have been included.

If the 2014–21 PSD, PPP and especially the SA failed to meet the needs of LPAs, did the 2022 revisions plug the gap?

2.2 NED guidance 2022

Many of the NE SA documents still available on 1 January 2022, were dated. They were unsuitable for use by LPAs and often unsuitable for consultants too.

On 7–14 January 2022, a raft of new SA guidance appeared on the NED website under the Natural England banner. Given that some of the pre-existing SA was seven or more years old, it should have meant a warm welcome by LPA planners and others. Would this have been merited after an open evaluation of the revisions? As will be shown, 'no' seems to be the answer.

With LPA staffing levels under strain, with competence and capacity to deal with biodiversity issues at a low ebb, in order to be useful, any new SA would need to be succinct: to set out what needs doing, to not require external data sources, and to be equally usable by planners and applicants alike.

Good SA would be unequivocal and serve as the basis for categorical acceptance or rejection of an application by planners. The revised 2022 SA met few of these precepts. As a result, it muddied, rather than clarified, the field for both applicants and planners.

PSD and PPP – 14 January 2022

Both PSD and PPP advice were updated in January 2022, along with the species guidance (SA) to which they made cross references.

The changes were not fundamental, and neither asked for more detail to be provided by developers. The revised 2022 SA, which the document introduced was less prescriptive and less informative, and subtly shifted the burdens and levels of evidence from applicant to LPA.

In the 2022 PSD (Natural England 2022), there were no differences from earlier versions until the short section on 'enhance biodiversity', which had a nuanced change: 'You should exercise your biodiversity duty' became the less confrontational 'to meet your biodiversity duty'. A similar subtle change followed, as 'You can help the developer achieve a net gain in biodiversity by following the National Planning Policy Framework' became 'achieve a net gain in biodiversity through good design, such as green roofs, street trees or sustainable drainage'.

The latter change brought in, generically, the idea of biodiversity net gain (BNG), and the web link took a reader to a general list of 'good things' to do. The context and nuance that would allow readers to understand what was required, or meant, were missing.

The 2022 PPP (Natural England 2022a) document set out the simplest of messages that developers might want when preparing a planning application: it outlined a set of steps for what biodiversity information might be needed to accompany a planning application. One key initial step was missing – one that is at the core of the PSD and is central to the SA: prior consultation. Nowhere in the PPP was the developer required to consult an LPA prior to submission, as the PPP only discussed the need to consult with an LPA where it is likely that an LPA might reject, or had rejected, a planning application.

The 2022 PSD (Natural England 2022), to be used by LPA planners started from the opposite position:

1. Discuss survey requirements with developers.
   Before you consider a planning proposal, you should discuss the survey requirements with the developer.

Put this way, it is a very odd approach. LPAs have enough to do, without going out of their way to pre-screen a development that has not yet come to them or when it is not being requested by the PPP. On that basis, it seems odder still for the PSD to then proceed (after some hyperlinks that are for the developer, not the LPA) to state that you (the LPA):

should ask for a survey if:

- there is suitable habitat on the site to support protected species
- it is likely that protected species are present and may be affected by the proposed development
- protected species are present but you're not sure if they'll be affected.

The 2022 PSD then continued in this tone – repeated largely in the new SA – to request the LPA to do things that the developer had been expected to do pre-2022. Both the PPP and the PSD ended up referring to the needs of protected species and the SA. If both need the SA, then it is important that the 2022 SA was clear, concise and unambiguous.

2.2.1 The January 2022 suite of Standing Advice guidance in a nutshell

The 2022 SA slightly widened the range of species/groups covered in the pre-2022 SA (Table 2.1) by separating pearl mussels from invertebrates and adding fungi and lichens to protected plants: not something that all botanists might agree on.

The pre-2022 SA was largely self-contained and covered advice for both the ecologists hired by planning applicants and the LPA planning staff.

Individual pre-2022 SA normally opened with a sense of certainty:

> This is Natural England's species standing advice for local planning authorities who need to assess planning applications that affect Badgers. This information can be used to *decide* what is needed for surveys and when assessing mitigation measures for Badgers.

That sense of certainty – including the use of information to decide – was emphasised, reinforced by the need for the developer's ecologist to explain where and why the SA, as set out, might not have been applied. That would be critical in helping an LPA decide on an application's suitability.

> Ecologists need to decide which survey and mitigation methods are right for the project they're working on. If this standing advice isn't followed, they'll have to include a statement with the planning application explaining why.

That unequivocally placed the burden of proof on the applicant.

In the January 2022 revision, the focus of the SA was switched round by 180 degrees, so that it was almost exclusively aimed at the LPA alone. At the same time, the 'burdens' were moved from developer onto the LPA – just as Michael Gove had suggested in 2018. As the title and opening line to each species in the 2022 SA (Natural England 2022b) stated:

> [Badgers]: advice for making planning decisions
>
> How to assess a planning application when there are [Badgers] on or near a proposed development site.

Table 2.1 Species/groups covered by NE standing advice pre- and post-2022.

≤2020 Title and last revision	14.1.2022 +amended or new title	Comments
Badger 28.3.2015	*	
Bats 28.2.2020	*	
Birds 28.3.2015	*	
Freshwater and migratory fish 28.3.2015	Fish	
	Freshwater Pearl Mussel	Split off from invertebrates
Great Crested Newts 12.11.2020	*	
Hazel Dormouse 28.3.2015	*	
Invertebrates 10.8.2015	*	
Natterjack Toad 12.5.2015	*	
Otter 5.4.2019	*	
Plants 23.4.2015	Protected plants, fungi and lichens	Fungi and lichens added
Reptiles 28.3.2015	*	
Water Vole 23.3.2015	*	
White-clawed Crayfish 9.10.2014	*	
Ancient woodland, ancient trees and veteran trees 5.11.2018	*	

The 'you' in previous versions of the SA aimed at the applicant's ecologist was dropped in the 2022 version. The focus or emphasis was very different, falling squarely on the shoulders of the LPA. After a short contents list, each new 2022 SA stated for a species:

> This is Natural England's 'standing advice' for [Badgers]. It is a material planning consideration for local planning authorities (LPAs). You should take this advice into account when making planning decisions. It forms part of a collection of standing advice for protected species.

The reference in the SA to the species as a material planning consideration was new and might appear at first sight to offer more weight to the considerations, but it seems to imply a watering down of cover. No longer was the SA to be used to *decide*. Instead, biodiversity and protected species as a material consideration meant (Planning Portal 2023):

> taken into account in deciding a planning application or on an appeal against a planning decision

This means that biodiversity and protected species were now part of a range of factors to be considered. What other issues were to be used, or taken as counterweights, is not clear. Instead, NE assumed that the advice was adequate because Natural England (2022a):

> Following this advice:
> - avoids the need to consult on the negative effects of planning applications on [Badgers] in most cases;
> - can help you make decisions on development proposals.

What else might help planners decide development proposals was left unsaid. Here, and elsewhere, the 'you' in 2022 was now the LPA planner. The text continued:

> You may need a qualified ecologist to advise you on the planning application and supporting evidence. You can find one using either the:
> - Chartered Institute of Ecology and Environment Management (CIEEM) directory
> - Environmental Data Services directory.

That was a significant switch because, in 2021 and earlier, PSD advice to local authorities had stated:

> Use an expert, such as your local authority ecologist, to help you apply the standing advice to planning decisions if you're not a wildlife specialist.

The change may well have reflected the realisation that LPA planners lack capacity and competency in biodiversity matters and that the LPA's previously employed specialist staff had been weeded out. It also opened up a financial can of worms for LPAs in the future because LPA finances are stretched, and resources have declined. It risked compromising proper scrutiny for those LPAs that chose not to/could not fully allocate resources to deal with biodiversity in planning – even though it was now unequivocally a material planning consideration. This is already a problem in Scotland (CIEEM 2022).

The 2022 set of SA included a short note on how a species or group is protected. For Badgers, for example, a list of offences is given, with the injunction that the LPA should consider if the developer (no longer called the applicant) had complied with the legal protection of Badgers.

In the pre-2022 SAs, the *developer's* ecologist was required to decide if surveys were needed, due to either:

- signs of species on or near the site;
- historical or distribution records.

In the 2022 SA, the *LPA* was told to *ask* for a survey if:

- Historical or distribution records suggest that [Badgers] are active in the area – you can search the National Biodiversity Network Atlas by species and location if the species is/was on or near the site.
- There is suitable habitat for sett building or foraging.

The effect was to place the onus onto the LPA, not the developer, at an early stage of the process. It also suddenly placed emphasis on pre-submission discussions at a time of staff shortage. In addition, the LPA was also required to check the competence and capacity of the developer's ecologist to undertake the necessary work. That was plainly switching the pressure onto the LPA in the SA, and away from the developer.

This suggests that the LPA was to do the initial legwork. That may well be an imposition that will only be honoured in principle, rather than practice, due to resource constraints shown by the capacity and competence limits previously noted by ALGE (2013, 2020; Snell and Oxford 2022) and CIEEM (2019a).

In the pre-2022 SAs, aimed at the developer's ecologist, there were detailed entries for required survey methods and survey effort. Both were absent from the 2022 revisions. As a result, it is uncertain how, or why, LPA planners in 2022 or beyond could determine the adequacy of anything that comes over their desk in regard to protected species. No additional sources of detailed information were cited. In addition, the Natural England (2022a) 2022 SA stated that it expected that the ecologist would follow BS 42020 (BSI 2013); quite which aspects of what is a complex document should be used, and how the planner would know, is left unstated. This is covered in Chapter 3 of this guide.

Given that the mitigation hierarchy used by NE (avoidance, mitigation and then compensation) is then invoked in subsequent sections, it was surprising that the term 'impact' is not used. Instead, the development proposal had to show how likely it is that the species might be affected by the proposed development work. Those two are subtly different. The clarity and expression for this part of the SA varied strongly between the SA texts for individual species. No simple reference point was provided in the SA to allow the LPA to assess the basis or reliability of claims of non-effect. It is this lack of a yardstick in all texts that makes the revised 2022 SA so weak and insubstantial.

Having said that the new 2022 batch of SA was written directly at the LPA planning audience, it seems odd that generic avoidance, mitigation and compensation measures were largely expressed and directed at the *developer*, with lists of simple actions. Pre-2022 SA was far more detailed and helpful to ecologist and LPA alike. Few of the new 2022 SA documents required the developer to demonstrate the efficacy or scale of actions in detail. For underqualified LPA planners seeking to understand quite what is going on, this may well be a place where qualified, well briefed, consultants may make a difference – at a price.

The last two sections of each new SA covered biodiversity enhancement and site management and monitoring. All referred by hyperlink to meeting the LPA's biodiversity duty under the NERC Act of 2006. Under biodiversity enhancement (Natural England 2022a), the SA stated that the LPA should suggest ways for the developer to:

- create new or enhanced habitats on the development site
- achieve a net gain in biodiversity through good design, such as green roofs, street trees or sustainable drainage.

There was no discussion about habitat suitability, size thresholds or any ecologically sensitive issues such as connectivity or fragmentation, which should be borne in mind. How such habitats should be chosen, or recognised, or the timescales involved before they become effective, was ignored. Similarly, the examples stated to fulfil net gain would apply to almost none of the species listed in the SA. These are fig leaves, not ways of achieving credible biodiversity gains.

As post-development assessments are critical to checking whether habitat alteration or net gain have been achieved, it is imperative that the guidance on this is clear, gives direction (what needs doing, for how long, and over what sort of area and whether it uses techniques that complement the original surveys) and can be audited. The 2022 SA stated:

Site management and monitoring
You should consider the need for site monitoring and management. These measures are likely to be needed by protected species licences.

For Badgers, the SA continued that the plan should:

- make sure sett building and foraging habitats are intact and still available in the long term;
- check that setts have not been interfered with after development, such as from increased human presence or vandalism.

This can include carrying out management works to habitats and additional survey work to check that mitigation measures are working as intended, followed by remedial work if needed.

None of this had any spatial or temporal element, nor any sense of requirement. Consideration of the need is not necessarily going to produce a mandated action; it is not obvious how net gain could possibly be evaluated and delivered if there is no necessary follow up. That is very much an elephant left standing in a room.

The LPA may well mandate certain works under planning conditions, but if relying on the vagueness of the SA, the LPA is left without any sense of how and where this would need doing. Given the recognised problems of competence and capacity limits of LPA staff, the risk was that follow up may not be as robust as might be expected or required.

The effect of such loose SA documents was to place the onus onto the LPA, not the developer. For example, at an early stage the LPA is required to check the competence and capacity of the developer's ecologist to undertake the necessary work. Depending on CIEEM competencies is not as clear as it may seem. That was plainly switching the pressure onto the LPA, not the developer.

With the emphasis firmly on the LPA to drive the data quality process under the 2022 SA, it is unfortunate that the guidance fails both sides. Quite how far the SA has moved backwards is shown in the Table 2.2, which compares pre-2022 with the most recent January 2022 version of the SA.

Table 2.2 The pre-2022 and 2022 SA compared. Titles used in standing advice given in italics.

Coverage and categories used in NE standing advice pre-2022 and in 14.1.2022 updates

Guidance source	Focus	Legal coverage	Survey methods and details	Survey methods	Survey effort required	Impacts assessed and stated	Mitigation hierarchy	Habitat enhancement	Post-development		References
SA pre-2022			*Decide if you need to survey*	*Survey methods*	*Survey effort required*	*Assess the Impacts*	*Mitigation and compensation methods*		*Management and site maintenance after development*	*Additional licensing information*	
Comments	Focus on consultants to read, and undertaken in order to supply LPA with full details. Aim to allow LPA to see what was done and why actions taken. Some entries for use by LPA.	Legal/licensing usually covered at the end of the SA section. Not normally specific part of species SA texts.	No formal objective sought or stated but normally assumed to be presence/absence. No formal reference to PEA desk or walkover surveys. Focus is on consultants to survey and gather data.	No formal references. Outlines details and methods for consultant use; often insufficient for LPA needs. Includes comments on survey effort.	Used in some (e.g., Badgers) but not others (e.g., Otters) so that LPA would struggle with consistency and expectations.	Asks for assessment of effects without any mitigation. Basic precursor to measure need for mitigation or compensation.	Most stress avoidance as the first step, then use mitigation. Compensation only if negative impacts remain. Details suitable mitigation or compensation in sparse outline – too little for LPA practical use for most species.	Rarely used and no formal entry.	Most cases recognise that post-development follow-up monitoring is required. 5+ years is the norm.	Licenses include disturbance, mitigation or survey, depending on species. Normally the last entry in the SA where it is used – infrequent.	No formal list; the few web links given crash on use.

(continued)

Table 2.2 The pre-2022 and 2022 SA compared. Titles used in standing advice given in italics. (*continued*)

Coverage and categories used in NE standing advice pre-2022 and in 14.1.2022 updates

Guidance source	Focus	Legal coverage	Survey methods and details	Impacts assessed and stated	Mitigation hierarchy	Habitat enhancement	Post-development	Planning and license conditions	References
January 14, 2022 SA		*How [ ] are protected*	*When to ask for a survey*	*Assess the effect of development on [ ]*	*Avoidance, mitigation and compensation measures*	*Enhance biodiversity*	*Post development monitoring*	*Planning and license conditions*	*References*
Comments	Focus on LPA not consultants. Few entries for consultants. Details are general, not prescriptive. No call to explain applicant's actions or methods, unlike pre-2022.	The second entry for all species/ groups. Basic law and offences listed. Unlike pre-2022, states species/ group as a material planning consideration.	New title and focus are on LPA having to ask; depends on LPA knowing about distributions and habitats. No formal entry for consultants. No reference to PEA by consultants. Reads as if LPA might do this. Given accepted capacity and competence issues in LPAs, a major problem.	No formal entry. Refers to distributional records and habitat – for LPA to check, not applicant. No method details, times or distances. Not suited for LPA or use as guidance by consultants. Insufficient detail for use.	Not given. LPA required to check consultant competence – an invidious position.	States that level of mitigation needed requires understanding effects of development; no use of term 'potential development'. No separate section covering impacts; actions to be taken listed as if they might occur.	Start with avoidance, then mitigation and finally compensation for most species. Far less detail than in pre-2022 SA documents. Tone is discretionary and spatial/ numeric details missing.	Refers to LPA duty under NERC 2006 Act. Generic suggestions of very limited value.	Simple actions listed, but normally no details on scale, frequency, duration or other parameters that would help LPAs.	States developer may need licence. LPA told to check if uses conditions and that these are satisfactory under license terms. Capacity/ competence issues in LPAs ignored.	Limited use of hyperlinks. No formal reference lists.

[ ] individual species
LPA = local planning authority; PEA = preliminary ecological assessment.

GUIDANCE AND INTERPRETATION | 41

If the new 2022 SA guidance is imprecise, then it is important that consultants follow, and apply, the best guidance that is available and apply the British Biodiversity Standard BS 42020 (2013) as expected by the January 2022 standing advice. This is covered in Chapter 3.

2.2.2 Scotland the brave?

Before reviewing BS 42020, it might help to look at the experience in Scotland, to see if the 2022 NE updates and approach were mirrored by the 2020 changes at NatureScot. Was NE alone in its problems, or were there common issues for government advisers in different jurisdictions?

NatureScot is responsible for conservation and landscape protection and policies in Scotland, and it provides planning advisory material for use in LPAs across Scotland. The legislative background is similar, and also different. Nonetheless, LPAs in Scotland are responsible for planning advice and evaluation.

In March 2020, NatureScot (2020a) provided an updated set of advisory material for protected species. Much of the introductory section in each of the species-specific guidance was similar to that in the 2022 NE documents, but it differs thereafter.

Many of the resource and skills problems faced by LPA staff in Scotland mirror those found in England, as noted when staff were asked about their competence and capacity to deliver NatureScot guidance in practice (CIEEM 2022).

2.2.2.1 Protected species advice in Scotland in summary

The preamble to the individual NatureScot (2020a) protected species standing advice (PSSA) states:

> This is standing advice to help planning applicants seeking permission for development that could affect [bats], and to assist planning officers and other regulators in their assessment of these applications. It avoids the need for us to advise on individual planning consultations in relation to [bats]. We will only provide further advice in exceptional circumstances that are not covered by this standing advice.

This means that the guidance needs to be backed up with clear definitions and sets of supporting technical references, something that NE failed to do in the 2022 SA revisions. The similarities and differences between the two sets of guidance are given in Table 2.3 below.

The contrast between the two is striking (see Table 2.4); the 2022 NE SA is vague, largely discretionary and puts the onus onto the LPA from the start. In contrast, the PSSA starts with the premise that it is for applicants, and they have to follow the guidance – and this includes multiple working web-links – and NatureScot will only provide further advice in exceptional circumstances. To state that, and to expect it to work, requires clear, multi-referenced, guidance; that is pretty much what an applicant and their ecologist get. While it may be hard work to make sure that all of

Table 2.3 The recent NatureScot (2020a) PSSA guidance and the January 2022 NE Species Advisory Guidance (SA) compared

Coverage and categories used in NE standing advice pre-2022 and in 14 January 2022 updates

Guidance source	Focus	Legal coverage	Survey methods and details		Impacts assessed and stated	Mitigation hierarchy	Habitat enhancement	Post-development		References
			Carrying out a [ ] survey	Reporting survey results				Protection plan	Licensing development works affecting [ ]	
March 2020 PSSA		Legal protection for [ ]			Protection plan	Measures to minimise impacts on [ ]				
Comments	Focus on applicants submitting to LPA what is needed. Clear focus on implementation and data needed. Initial PSSA supported by detailed legal and methodological guidance. Includes how to set out reporting survey results.	Like the SA, Legal is the first entry. In the PSSA. Explains what and how [ ] are protected and what are offences. When a development could affect [ ], it explains context of possible concern plus data sources and cross references.	Sets out basis, techniques and caveats and provides cross references for detailed best practices. Includes notes on data sources and age. Distances and timing given. No reference to PEA, as focus is on protected species alone.	Provides detailed contents of report structures. More detailed references given. If species could be affected, then pro forma for development plan provided.	Includes details of plan contents, minimisation of impacts and licensing requirements and detailed references.	Starts with avoidance details, mitigation details and, where necessary, compensation. References cited for more detailed coverage. Need for post-submission surveys stated so as to be up to date on final planning and implementation of project.	Not noted separately, and included as part of any compensation package.	Protection plan includes an obligatory monitoring element in order to assess effects of applying mitigation hierarchy and adjust as needs	If impacts unavoidable, then specifications for works and terms set out for each species. NatureScot license requires clear alternatives and protocols. References for methods provided.	Hyperlinked references provided in each section. Allows LPA to follow needs through.

January 14, 2022, SA	How l l are protected	When to ask for a survey	Survey methods	Survey effort required	Assess the effect of development on [ ]	Avoidance, mitigation and compensation measures	Enhance biodiversity	Post development monitoring	Planning and License conditions	References
Comments	Focus on LPA not consultants. Few entries for consultants. Details general, not prescriptive. No call to explain applicant's actions or methods, unlike pre-2022.	The first entry for all species/ groups. Basic law and offences listed. Unlike pre-2022, states species/ group as a material planning consideration.	New title and focus are on LPA having to ask; depends on PEA by consultants knowing about distributions and habitats. No formal entry for consultants. No reference to PEA by consultants. Reads as if LPA might do this. Given accepted capacity and competence issues in LPAs, it is a major problem.	No formal entry. Refers to distributional records and habitat – for LPA to check, not applicant. No method details, times or distances. Not suited for LPA or use as guidance by consultants. Insufficient detail for use.	Not given. LPA required to check consultant competence – an invidious position.	States that level of mitigation needed requires understanding effects of development; no use of term potential development. No separate section covering impacts; actions to be taken listed as if they might occur.	Start with avoidance, then mitigation and finally compensation for most species. Far less detail than in pre-2022 SA documents. Tone is discretionary and spatial/ numeric details missing.	Refers to LPA duty under NERC 2006 Act. Generic suggestions of very limited value.	Simple actions listed, but normally no details on scale, frequency, duration or other parameters that would help LPAs.	States developer may need licence. LPA told to check if uses conditions and that these are satisfactory under license terms. Capacity and competence issues in LPAs ignored.	Limited use of hyperlinks. No formal reference lists.

Titles used in standing advice given in italics. PSSA = protected species standing advice; NE= Natural England; LPA = local planning authority; PEA = preliminary ecological assessment.

Table 2.4 Comparisons of the new advisory guidance for protected species – in this example the Badger – in NE and NS.

Category and question	England SA 2022	Scotland PSSA 2020	Comments
Who is the SA/PSSA aimed at?	LPA	Applicant	Polar opposite starting points
Is the SA/PSSA meant to preclude the need to consult with NE/NS?	Yes	Yes, in principle	If there is a threat to a protected species and a licence will be needed, applicants must still contact NS
Is the material in the SA/PSSA self-contained/ well enough referenced to do this?	No	Yes	PSSA guidance backed up by multiple web links to detailed topic guidance
BADGER			
Are offences under relevant Act listed?	Yes	Yes	SA totally inadequate. PSSA references provide detailed guidance
Is the need for licensing flagged up early on?	Yes	Yes	
Are ways of noting potential effects on Badgers from a development stated clearly?	No. The SA assumes that LPA will look at either the NBN or know of habitat suitability	Refers to NBN, Local Record Centres and Mammal Society Atlas plus regional reference notes. Habitat types stated	PSSA provides far more initial detail, then supportive material in web links
How far should surveyors look?	No distance given	Initially <100 m, overall, within 1 km in linked guidance	SA fails to note distances. PSSA web links to detailed references
Are enough details given to assess if survey efforts are suitable?	No. No details offered	Yes. Initial PSSA outlines methods, and web links detail techniques and frequency	SA totally inadequate. PSSA references provide detailed guidance
Can LPA use SA/PSSA to see limitations in any submission?	No	Yes, if they refer to supplementary NS documents	No basis for LPA to spot errors/limitations in SA-based data
Does the LPA have to check suitability of ecologist used?	Yes, based on CIEEM competencies	No, that is the role of the applicant to submit as part of the application	Competencies not enough as licence may also needed
Is it clear how old data can be and still be used?	2 years	2 years, but needs to be close to start of project to check for changes after application submitted	PSSA requires new surveys if older than 2 years or delays occur
Does the guidance allow an LPA to tell if an applicant did the right surveys, when where and how?	No	Yes, PSSA outlines basics and linked pages tell details of methods and problems with them	SA totally inadequate; PSSA and links provides detailed guidance, including regional issues to consider

Does the guidance show what a report should contain?	No	Yes, and headings are given	PSSA is about quality control; SA ignores quality
Does the guidance call for a formal protection plan?	No	Yes, and it outlines the contents and links to a detailed plan guidance document	The SA suggests that applicants 'should consider' the need for a site management and monitoring plan and mitigation proposals
Is the mitigation hierarchy invoked to minimise impacts?	Yes, SA calls for proposals that will avoid, mitigate and, if necessary, compensate	Yes, covered in the protection plan, but lists actions and distances under avoidance, mitigation and compensation if needed	SA omits distances and details that PSSA contains that inform planners and applicants
Is compensation realistically covered?	No. SA states 'could include creating artificial setts as a compensation measure'. No details provided	Artificial setts not in PSSA compensation section, which instead refers to foraging habitat enhancement to offset loss of feeding areas	PSSA details artificial setts as a 'least preferred option' and covers in detail licensing works instead
How is licensing addressed?	SA refers to the possibility of licensing, in which case mitigation/compensation may be conditioned. LPA needs to be sure that NE will issue a license	PSSA provides detailed links to guides on license applications and artificial sett creation. Would-be developer required to contact NS licensing team and include outcomes as part of application	SA offers minimal detail on why and how licensing would be required; PSSA provides a detailed resume and web links and states the direct need to contact NS licensing team as part of an application
Is enhancement a separate issue?	SA refers generically in a separate section to creating or enhancing site habitats, or adding green roofs, sustainable drainage systems or street trees	Not a category covered formally in its own section, but is covered under compensation	SA categories are too generic to be helpful and not species-specific enough. Most examples are not suitable for protected species
Site management and monitoring	SA refers to the possible need for these via a site management and monitoring plan that ensures long-term habitat continuity and sett survival	In the PSSA and linked guidance, a protection plan sits centre stage where Badgers might be affected by a proposal. Strong contrast between SA and PSSA details	SA presents very limited guidance in contrast to the PSSA guidance and detailed web links

SA = standing advice; PSSA = protected species standing advice; LPA = local planning authority; NE = Natural England; NS = NatureScot; NBN = National Biodiversity Network; CIEEM = Chartered Institute of Ecological and Environmental Management

the cited links are understood, that is for the applicant to do. That is a far cry from the equivocal 2022 NE SA and the pressure placed on LPAs in England.

It may help to compare and contrast the two sets of guidance by looking at a sample species: the Badger. The Badger is a protected species in both countries, but the detailed level of guidance varies significantly between them.

Even with a sound set of PSSA guidance to hand, there is no guarantee that LPAs in Scotland are set up to fully evaluate what comes over their desks. A recurrent theme of English-based ALGE surveys has been the lack of capacity (staff time) and competence (expertise) to tease apart planning applications. Without these, no matter how good, or bad, the applications are, they cannot be properly evaluated (CIEEM 2019a).

CIEEM, in association with ALGE (CIEEM 2022), reported on a survey of these attributes in Scottish LPAs in 2021. A total of 22% of the 36 respondents stated they had no current ecological resource or expertise available to them, and 31% reported a lack of capacity. Overall, 84% of respondents reckoned that they could only properly address the need for positive effects for biodiversity with more resources, greater capacity or reduced workload. Only 11% of LPAs had adequate resources and capacity. Put together, that is not an encouraging position for biodiversity in the planning system in Scotland.

Even where there was capacity and competence – which was limited – two thirds of LPAs reported that lack of enforcement staff was a major risk to their ability to deliver on planning requirements and ensure positive effects for biodiversity. CIEEM (2022) indicated that without an adequately trained and resourced set of environmental planners and ecologists in each Scottish LPA, then it will be impossible to deliver positive effects for biodiversity across the country.

2.3 The Partnership for Biodiversity in Planning

The disquiet with the old NE documents was reflected by the founding of the Partnership for Biodiversity in Planning (PBP) in 2018. It describes itself as:

> an alliance of 19 organisations representing the conservation, planning and development sectors, who are working together to simplify, streamline and improve the consideration of biodiversity in the UK planning process.

The PBP includes the major species conservation groups for birds, mammals, herptiles and invertebrates as well as professional planning organisations and individual LPA planners and academics. The PBP developed the WAC as part of an attempt to:

> help consider protected and priority species earlier in the UK planning process and encourage building projects to deliver a net gain in biodiversity.

A key part of this was the development of its web-based screening planning tool – the wildlife assessment check (WAC) – which was built:

> to offer householders and small to medium scale developers a simple first check to see whether a potential development project requires expert ecological advice.

The drive behind the development of the web-based screening tool was the recognition that it was the statutory duty for all public authorities to have regard for biodiversity conservation, and the belief that the planning system offers an opportunity to reverse biodiversity loss through promoting wildlife conservation and restoration. To do this requires suitable documentation to be generally available, something the NE documentation was seen as not providing. PBP was aware of the shortfall of ecologists in LPAs noted by ALGE (2013) and others.

For PBP, the risk was heightened by the fact that small-scale developers were often unaware that their planning applications were incomplete because they had not sought expert ecological advice as a part of their application. PBP saw that the outcome was delay and additional knock-on costs when unforeseen ecological surveys were needed by developers. As the first check, or screening, WAC was seen as the way out, identifying whether certain protected and priority wildlife species, as well as special designated sites, might be affected by a building project. It could also indicate whether an expert ecologist might be required by the applicant in advance of submitting an application.

The WAC was pitched as a triple win. For planners, it would help to meet statutory obligations by considering ecology in planning applications. For developers, early consideration in the planning process could reduce costs, and by making developers consider ecology in projects it would be expected to improve consistency and standards. Financed by charitable grant aid until early 2020, PBP's WAC tool is now subject to a maintenance and retention programme, rather than ongoing development.

The WAC is in fact two different sorts of offerings: the screening service outlined above, and a set of more detailed species survey guidance that would need to be followed by consultants if problematic species were identified or if LPA planners wanted to understand in more detail the suitability of what was being submitted to them. In principle, the latter is complementary to the SA. That the PBP felt the need to do their own version of guidance for LPAs tells something about how the SA was viewed by LPAs and others in the consortium.

2.3.1 The Wildlife Assessment Check

The web-based WAC screening check has four simple steps or checks:

- Check 1. Project location: click on the map to locate the project.
- Check 2. Type of development project: click on whether it is a large (schedule 1 or 2) development, smaller development or an individual household project.
- Check 3. Site and works context: click on relevant boxes to indicate the type of natural habitats that may be affected (e.g., waterbodies, trees, hedgerows, coastal areas) and works that are being carried out.
- Check 4. Further details: click on relevant boxes to indicate more detailed information about the habitats and types of proposed work.

One further click, and the user is presented with:

- Results and summary report: a summary of the project information and ecological advice is produced.

For PBP, once complete, the results page of the screening check would indicate whether the project site would be likely to require professional ecological advice for any protected and priority species, or statutory designated areas that may be impacted by the development, before making a planning application or submitting a permitted development. The short 'summary report' produced could be given to a consultant ecologist, and submitted with any planning application, and additional material gathered as a result of the check.

By its own admission, the WAC screening option is less powerful than it claims, and its limits are noted in the first few screens of the website:

> The tool uses national species maps and triggers that do not always pick up **local** species data. In addition, although certain natural **habitats** are associated to protected and priority species, the tool does not hold information on **'priority habitats'** (such as rivers, hedgerows and ancient woodlands) or Local Wildlife Sites that may need to be considered in terms of ecological impact. So we would always advise that the ecological consultant seeks additional information from the local environmental records centre and Local Wildlife Groups, as well as consider conducting a Preliminary Ecological Appraisal.

That first set of caveats should be read as a health warning, as the screening is far from complete; many datasets may never get to national level datasets or even local record centres. A second group of caveats was also posted before would-be users entered into the screening site:

> **NOTE:** It is important for users to note that the Wildlife Assessment Check is for **guidance only**. It is not designed to replace the judgement of a qualified professional ecologist about the potential wildlife impact of a development project. It has been developed for local authorities who have more limited in-house ecological capacity. Applicants should consult their local authority ecologist where they are present, to ensure that a proposed development does not require an ecological appraisal.

Of course, having been set up to deal with the shortfall of local authority ecologists, asking an LPA ecologist for comments seems rather counter-intuitive. That was an issue in 2013 (ALGE 2013) and remains so a decade later (Snell and Oxford 2022).

2.3.2 Species Guidance Notes

The second main value of the PBP comes from the species guidance notes (SGN) – intended primarily for use by consultants, but also as a check for LPA use. The notes were produced by the species-oriented NGO members of the PBP.

Are the SGN a straightforward replacement for the SA? The simple answer is no; there is a mismatch in species/groups covered by the SA and the SGN. Ancient woods and veteran trees and freshwater fish are missing from the SGN, and Natterjack Toad and crayfish are covered more generally than specifically in the SGN.

For those species/groups that are covered, the basic issue is whether the SGNs are an improvement on the pre-2022 and post-2022 SA; whether they are self-standing or rely on a combination of ecological expertise at planner or advisory levels; and whether they require the user to turn to the references for detail and explanation. Doing so would negate the role of that individual species SGN. If any of these caveats apply, that would also mean that the LPA had come full circle – needing the expertise to understand the guidance that the existence of good guidance was meant to preclude.

Contents and structure of the SGN

There is a standard formula for each species/group, starting with legal protection and ending with references. As the SGN covers all four jurisdictions within Great Britain and Northern Ireland, each species has separate entries for legal protection – as this varies between countries. Licensing also has jurisdiction-specific notes. The basic contents for each species/group include:

>Legal protection: the national legislation that provides protection.

>Survey information: includes advice regarding suitable seasons to conduct surveys, and survey methods for that species/species group.

>Mitigation: outlines measures to be taken to avoid, mitigate (lessen) or compensate for negative impacts of development on species.

>Habitat enhancement: measures to create or improve a habitat to support that species/species group.

>Post-development monitoring: how that species/species group should be monitored once a development is constructed.

>Licensing: outlines whether licences are required in each country before any works can be undertaken that may negatively impact a species.

>References: key documentation with more detailed advice regarding that species/species group.

The headings overlap in general terms with those in the SA, apart from the absence of a formal impact assessment statement and the provision of a references section.

There is no impact assessment section. This would be needed to confirm the scale and location for mitigation and compensation. As anyone using the SGN entries may well need to find further details so as to be sure of what is/was needed, references are important; these mark a major improvement over their omission from the SA format.

2.3.2.1 Using the SGN format in practice

In purely appearance terms, the SGN scores well: the text is clearly set out and covers its topics in a series of short sentences or paragraphs rather than terse single line entries common in the SA. The use of a colour coded survey calendar allows the presentation of a lot of data in a concise form.

While simple presentation helps, it is critical that the details needed for each section are also clear. That way, the frequency, duration, distance and timings of visits and techniques should be readily apparent, and an LPA planner should be able to see what is needed or expected for any species/species group.

In the Appendix, detailed reviews of how well the SGN worked, when compared with the SA and standard species references are provided. For convenience, summaries of Badgers, reptiles and birds are given here, along with comparisons of the NE SA and SGN.

SGN and Badgers

The SGN reads as an introductory guide, framing the context, but short on details. If the objective is to ensure or promote standards, then the lack of detailed timings or distances, and the absence of limitations discussions, means that an LPA planner might well be unsure what is required, or why something was done, or if materials provided were adequate. Why the requirements of a full preliminary ecological appraisal (PEA) are left out is uncertain; desk-based distributional records may inform the scope and scale of walkover surveys. The absence of an impacts section seems surprising; mitigation is an activity relating to impacts. If these are not formally stated, then the SGN is procedurally deficient.

Although presentationally good, the absence of explanations and details means that a reader may well be forced to the secondary reference sources, rather than finding them in the SGN.

Reptiles and the SGN

The guidance ticks many of the boxes in outline, but misses several basic elements, such as the desk part of a PEA, detailed temperature issues and time of day, both of which affect detectability. Like all other SGN, the absence of an impacts section – linked to population estimates – limits the viability of assessing, and then monitoring, impacts.

Birds and the SGN

Birds are a complex group of species, and are a challenge to summarise for planning cases. The SGN focus on buildings and trees is an over-simplification of what is a complex reality in the wider ecological world. Even here, the guidance is too generic and caveated in a needlessly complex way. Limitations are inconsistently cited, and objectives with linked survey effort are covered poorly. Windfarms get mentioned in references but not in the text. As many LPA public inquiries across the UK involve windfarms, and may yet do so again in England, their non-coverage seems an important oversight.

The recognition that the SGN bird guidance was weak was confirmed by the provision of a separate subset of guidance for birds, produced outside of the SGN by members of the PBP.

2.3.2.2 Birds and bird survey guidelines: a complement to the SGN?

As birds feature in almost all planning applications, it may help to consider what the SGN did not do, and why another partial version of guidance was produced before

GUIDANCE AND INTERPRETATION | 51

the SGN was completed. The WAC/SGN was developed by a consortium of partners, including ALGE, RSPB and the Wildlife Trust as a tool for LPAs. Yet, less than a year after its launch, an additional web-based bird survey resource was made public in May 2021. The Bird Survey Guidelines (BSG 2023) was the product of ALGE, British Trust for Ornithology (BTO), RSPB and a range of commercial consultancies. Sharing many of the sponsors and personnel of the WAC/SGN, it was:

> Intended that these guidelines will provide an approach that can be consistently implemented across the ecological consultancy industry, and it is therefore, applicable to the collection of ornithological data in both terrestrial and freshwater habitats.

At its launch in 2021, it only covered the breeding bird season, as

> there is an absence of a clear methodology for undertaking ornithological surveys in the bird breeding season. It is the purpose of these guidelines, to address this gap.

Given the overlap in membership, and the recency of the SGN, that was hardly a ringing endorsement for their earlier, limited, work under the WAC/SGN banner. The guide was:

> intended for use by ornithologists and ecological consultants who engage in bird surveys for the purpose of assessing ecological impacts including Ecological Impact Assessments (EcIA) and Environmental Impact Assessments (EIA).

Given the potential overlap or confusion between two very recent and overlapping products, if used by an uncertain LPA planner, was the BSG any better?

The BSG started from the premise that birds should always be scoped in as a species group to be surveyed, unless there is clear reason not to. That is more than a cursory visit or two. For its lead proponent (Hounsome 2021), this meant at least six visits to an area, largely regardless of site type.

Unlike the SGN, there is emphasis on desk as well as field studies in the first part of the PEA/EcIA stage. That is a positive step, although Hounsome (2021) noted that Local Record Centre data were often highly inadequate.

Users of the methods set out in the BSG were seeking:

> to measure in what way, if at all, a given area is important to avian diversity, what species could potentially be breeding and, therefore, what the impacts of the project or development being considered are likely to be. When gathering survey information, be mindful of the Precautionary Principle as defined by the Chartered Institute of Ecology and Environmental Management (CIEEM):

> The evaluation of significant effects should always be based on the best available scientific evidence. If sufficient information is not available further survey or additional research may be required. In cases of reasonable doubt,

where it is not possible to robustly justify a conclusion of no significant effect, a significant effect should be assumed. Where uncertainty exists, it must be acknowledged in the EcIA.

Unlike both the SGN and the SA, the BSG set out in detail the first scoping stage and the important use of data sources (Fig. 2.1).

Assuming that surveys were to be required, a reader is taken to the survey methods section where the basic expectations are provided. There is recognition of both generic methods and species-specific requirements, with these referenced.

Taking SGN category headings for ease of comparison, the level of survey effort is summarised, again with caveats and references for individual species. Unlike SGN, there is no division into trees or buildings. A minimum of six visits spread between mid-March and early July is recommended. There is discussion of exact timings. BSG recommends that surveys for breeding birds should be carried out approximately between half an hour before sunrise and mid-morning (10–11 am, with some regional variation). The start time differs by 1.5 hours from SGN, and the end time by an hour. As species vary in their detectability throughout the day, BSG requires that at least one of the six or more visits should be in the evening (i.e., during the last few hours of the day, and extending beyond sunset for at least one hour) to deal with species not readily recorded by conventional surveys early in the morning. Some species call into the dusk and after dark, which can be difficult to detect during the day. BSG sets out clear limitations affecting surveys and their reliability Thus there are significant differences between the two sets of advice: SGN and BSG. Of the two, BSG seems more realistic.

The BSG is under continuous development, and will provide far more details than the SGN.

Bird advice and the LPA planner

Planners are faced with a dilemma. The SGN guidance is quite generic, and omits full coverage of species and seems over-focused on a few habitat types. The guidance from BSG by contrast is more readily understood, more applicable and better referenced, starting with the premise that it is both the site and how it is used, and the immediate area around it, that needs to be understood to fully appreciate potential impacts (as suggested also by the introduction to the SA). In 2023, BSG expanded the guidance to cover non-breeding periods, providing a better, more readily understood source of guidance. How an LPA planner can reconcile basic guidance contradictions between two sets of guidance sharing common advisory panels is not obvious. Which of the two sets of times is right, for example? Reference to the SA will not help decide. Reference to Gilbert et al. (1996) or Bibby et al. (2002) would.

Can an LPA take the SA and SGN guidelines and determine cases? Without external guidance or resources, planners will struggle to know whether they have been provided with suitable materials, or how to interpret them; that is a problem at a practical level.

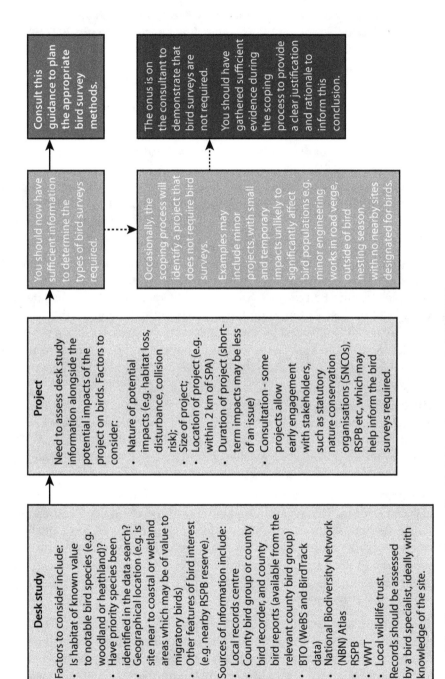

Figure 2.1 The process of evaluating bird risk and survey needs (BSG 2021).

2.3.3 The NE standing advice and SGN compared

For a long while, the only formal source of advice for planners on biodiversity matters was the NE SA given by NE on its section of the NED website. As noted, this has some serious limitations. From the LPA planners' perspective, the pre-2022 SA was not up to scratch, which led to the PBP and the WAC and its advisory offspring, the SGN. To what extent is the PBP's SGN an improvement, and where could changes be made to both to maximise the value to their client market?

Before comparing them, it might help to step back a little, and ask what does a planner or someone wanting to understand a planning application need?

The first step would be a set of simple, clear, concise headings that cover all that is required. Both the SA and SGN use headings that seem to promise most of the basics, but not all. When a planner is being asked to look at the ecological component of a planning application, they might reasonably expect a basic set of information for each protected group, or species, that includes a number of headings not found in the SA or SGN:

- The area covered: the actual site and site boundaries, and wider zone of influence. This is normally 2 km (CIEEM 2019a), and would include, for example, satellite Badger sets, areas home to species that use the site and areas used by species found on the site. For noise, light and other factors, these areas can be affected by direct or secondary impacts from a potential development. From a species' standpoint, both would be material considerations. Neither the SA nor SGN mention these consistently in their summary coverages.
- The objectives: clearly stated objectives help understanding what the data were collected for, and what they were not. That helps in assessing if they are fit for purpose, and what that purpose was. In the bird-specific BSG Hounsome (2021) was very clear that objectives need stating for bird surveys. These are omitted from both the SA and SGN.
- The method or standard used: every survey should have a reference that an LPA can refer to. This should be stated. For those using the SA, there is no stated method, so this means that an LPA planner is placed in a quandary when seeking to understand what actually took place, and whether any actual standard was used or applied. For the SGN, a range of possible references is provided, but it is uncertain which formed the core or key references, so clarity is hard to establish.

In their EcIA/EIA, many developers state that they 'followed Guide X'. There is a difference between followed and applied. The former is indicative, and the latter is much more categorical. From evidence viewed when working as an Expert at Planning Inquiries, 'followed' is usually a sign of loose adherence, with the variation rarely, if ever, stated. It is normally up to a reviewer to seek to see how far guidance, or standards were applied by the applicant.

As determining planning permission requires a validated evidence base, it would be better to state what was and what was not done in 'following' – as is required by

BS 42020 (BSI 2013), the British biodiversity standard (see Chapter 3), and how this influences the reliability of the data collected and presented. That would of course be expected by the CIEEM code of ethics, which requires use of and adherence to BS 42020, to which many consultants are bound; this often seems to be forgotten. It would be helpful if SA and SGN both stressed issues about the quality of material to be collected and provided.

Among the issues regularly missing in both of the SA and SGN are:

- Recency. In principle, all data for a site are helpful. Runs of data are part of the site scene-setting. However, if the objective of an EcIA or EIA is to understand current and future impacts, then it is important that recent data are used (CIEEM 2019b). Using old data alone is an insecure basis for impact assessment, especially if their collection context is not readily documented. However, both data age and context are frequently discounted as issues in assessments. Neither the SA nor SGN lay emphasis on these fundamental limitations. Some SA, such as for the Hazel Dormouse, clearly state the risks of using old data, while most ignore it.
- Limitations. Data are as good as the circumstances under which they were collected. Neither the SA nor SGN is consistent in identifying limitations. Normally, ecological guidance for data collection stresses the need to state limits/variations (BSI 2013; CIEEM 2019a), yet this is not consistently covered in either SA or SGN. If taking the SA at face value, then for most species/ groups there would be no formal need to cover limitations; that is unwise.

2.3.3.1 A tabular comparison of SA and SGN guidance documents

If the objective of the SA and SGN guidance is to guarantee that developers and their ecologists will collect suitable data, then it would be expected that:

- the guidance provides enough help to ensure this;
- that the guidance uses clear references;
- that the guidance is overt about the implications of limiting factors, such as weather, time of year, disturbance and the number of visits needed to fulfil stated objectives.

It will be seen in Table 2.5 and the Appendix that these are largely ignored.

In addition, if PBP were to have decided that the SA materials were inadequate, then it would be expected that it would at least cover the same species groups, and perhaps extend these if it was thought helpful. The SGN omits three groups/species in the SA list: freshwater fish, ancient woods and veteran trees, and lumps Natterjack Toads in with amphibians, rather than treat them separately. While SGN does extend the range of species beyond the short list covered in the SA, they almost all share the same set of limitations shown in the following tables: inconsistent levels of details in survey methodologies so that an LPA cannot reliably tell what is needed, absence of desk survey data, poor discussions of limitations, if given, and varying levels of summary of adherence to cited references. The result is that LPA or other users cannot readily get the working precis needed from either the SA or SGN (see Appendix).

For groups such as invertebrates or rare plants, which cover hundreds of species, it might not be unreasonable for the SA or SGN to fail to spell out individual specific survey protocols or methods. Both SGN and SA plant entries rely on weblinks to non-functional sites.

Even where it is recognised that there are detailed group or species guides, such as for birds, the failure to provide an overview and precis of the sorts of materials available, and their use, is surprising. For birds, the almost total absence of any coverage of methods for passage and wintering communities, habitats and sites, for which Britain has a wide array of internationally important sites (https://jncc.gov.uk/our-work/uk-protected-areas/), is an unexpected gap.

2.3.4 What the 2022 SA and 2020 SGN do and do not do well

Table 2.5 summarises the main category head points for the SA and SGN. The first issue is legal coverage. As the SA list covers protected species, it is unsurprising that licensing and legal protection are covered within the guide. The SGN typically gives as much, if not more, space to legal issues and licensing in the four jurisdictions than to survey methods.

The first question that any guidance should consider is whether there is any need to undertake a detailed survey. This relies on a basic initial desk-based data assessment and also a walkover survey: a PEA (CIEEM 2019a). The SA includes a formal note to decide if there is a need to survey – and includes a desk survey as part of this. The SGN does not. In the SGN, if a species is worthy of coverage in a planning application, then it needs a field survey – but is not, apparently, worthy of a desk survey. This seems a strange omission. It also would not fit with most LPA validation checklists (VC).

Survey methods and details are both covered in the SA and SGN, although with different terms. As the SA is often very concise, the absence of details and supporting information – and references – means that the SA rarely provides enough guidance for LPA use. The SGN provides variable levels of detail, rarely enough for effective use in planning assessment. Survey effort and survey calendar are standard parts of the SGN, although sometimes disagreeing with the cited standard references. The SA omits any summary graphics.

One of the key objectives in a planning submission, from an ecological standpoint at least, is to establish whether there are potential impacts. The SA is clear on what an impact is. It almost always offers some sort of short guiding list of what the main impact categories would be. Both the heading and the categories are missing from the SGN.

If a key pillar, impacts, is formally missed by the SGN, then it is important that it recognises the gradation of the mitigation hierarchy, from avoidance to compensation, in order to minimise any impacts. Mitigation options in the SA are few and usually

Table 2.5 Comparisons of main headings and contents in the NE and WAC SGN species guidance notes, and some of the issues remaining

Guidance Source	Legal coverage	Survey or not?	Survey methods and details	Impacts assessed and stated	Mitigation hierarchy	Habitat enhancement	Post-development	References
SA		Decide if you need to survey	Survey methods	Survey effort required	Assess the impacts	Mitigation and compensation methods		Management and site maintenance after development	Additional licensing information
Comments	Legal coverage at the start and licensing at the end of the SA section.	No formal objective sought or stated, but normally assumed to be presence/absence. No formal reference to PEA desk or walkover surveys.	No formal references. Details outline and methods can be too sketchy for LPA needs. Includes comments on survey effort.	Used in some (e.g., Badgers) but not others (e.g., Otter) so that LPA would struggle with consistency and expectations.	No survey calendar: most texts refer to dates.	Asks for assessment of effects without any mitigation. Basic precursor to measure need for mitigation or compensation.	Most stress avoidance as the first step, then mitigation. Compensation only if negative impacts remain. Details suitable mitigation or compensation in sparse outline; too little for LPA use for most species.	Rarely used.	Most cases recognise that post-development follow-up monitoring is required. Five or more years is the norm.	Licenses include disturbance, mitigation, or survey, depending on species. Normally the last entry in the SA.	No formal list; the few web links given crash on use.

(continued)

Table 2.5 Comparisons of main headings and contents in the NE and WAC SGN species guidance notes, and some of the issues remaining (*continued*)

Guidance Source	Legal coverage	Survey or not?	Survey methods and details			Impacts assessed and stated	Mitigation hierarchy	Habitat enhancement	Post-development	Licensing	References
			Survey information	Survey effort	Survey calendar						
SGN	*Legal protection*	*Survey information*	*Survey information*	*Survey effort*	*Survey calendar*		*Mitigation*	*Habitat enhancement*	*Post-development monitoring*	*Licensing*	*References*
Comments	The first entry for all species/groups. Separate for each jurisdiction.	No formal entry.	Normally (but not bats) there is a preliminary walkover survey, but no desk survey asked for. Helps decide if more survey work needed.	Gives both preliminary and further survey methods; not always in sufficient detail for LPA needs.	Outlines level of inputs required; not always in sufficient detail for LPA to evaluate claims.	Outlines optimal and sub-optimal times for survey; not necessarily explained.	No separate section covering impacts; actions to be taken listed if they might occur.	Start with avoidance, then mitigation and finally compensation for most species.	Generic activities listed	Simple actions listed, but normally no details on scale, frequency or others that would help LPAs.	Required entry for some species (e.g., Great Crested Newt, Hazel Dormouse)	Lists given, no obvious way of ranking priority

Category title (in italics) in order of listing in each advisory guide. NE= Natural England; WAC = Wildlife Assessment Check; SGN = species guidance notes; SA = standing advice; PEA = preliminary ecological assessment; LPA = local planning authority.

lack any spatial dimensions or review of time scales required. The SGN is highly variable on both of these. Habitat enhancement is not a heading used by the SA.

For the LPA hoping for clear, consistent guidance, neither the SA nor SGN provides a robust, consistent model on how to approach assessments coming over the desk. If a planner is concerned about checking whether what is claimed as a standard practice actually is, then being able to check through the key references is important. The SA cites almost none, and of these half refer to non-existent webpages. The WAC shares some of these broken links too. Where the SA was under-referenced, the SGN veers in the opposite direction. While that is not a critical sin, there is no guidance on core or secondary importance. As some of the references are long, detailed, or make assumptions of biological knowledge, that is a problem for time-strapped and competence-limited LPA staff.

The level of inconsistency between the sets of guidance is reviewed in Table 2.5 for the range of species or groups held in common.

2.3.5 Comparing and contrasting the SA and SGN at the species advisory level: key issues

In order for an LPA to fully evaluate a planning application, a planner needs to understand what took place, when, where and how widely the surveys addressed direct and indirect issues. A planner must also grasp the methods used (and if these were standard and if not then why not (see BS 42020 in Chapter 3)), the limitations of thereof and the likely impacts, as well as how to deal with these, evaluating the impacts and being able to see how these night be mitigated. Of course, if these were clear and concise, stating the necessary post-development monitoring protocols and evaluation would be help in the acceptance or rejection of a proposal. How well these are covered and how they compare in the SA and SGN is summarised for Badgers, reptiles and bats. Full comparisons are given in the Appendix for 10 species/groups to which both sets of guidance apply.

It would be assumed that the SA and SGN would be very similar – especially where the same basic sources would appear to have been consulted. This was not always the case.

Badgers

For a wide-ranging species such as the Badger, it is critical that distances away from primary setts are covered, as well as in the immediate 50 m, as any development might affect core, or marginal areas used by Badgers in the area; finding satellite setts and assessing how far the sett's occupants range is critical to understanding direct, indirect and cumulative impacts. Neither the SA nor the SGN address these fully in either initial surveys, impacts or mitigation. Both vary on advice about the best times of the year to view setts, as do details on the best period for bait-marking studies. Understanding the frequency of checks required for bait marking or sett use seems a basic requirement. Neither is fully addressed by either the SA or SGN. The duration for the placement of camera traps advocated by the SGN is similarly vague.

The SA list of impact issues and the linked mitigation options is clearer than the SGN, which assumes there might or might not be impacts and provides no list of what to consider. Having a list to consider seems prudent as the basis for mitigation. Only the SGN formalised the post-development monitoring requirement and duration. Any LPA concerned with checking issues in the SA would need to go outside of the guidance, but there are no references to consult. While the SGN has a range of references, which is the right one(s) is hard to discern; providing clarity and guidance – which is absent from the SA – was the part of the reason behind the WAC and SGN.

The lack of discussion over limitations, problems with spatial and temporal aspects of surveys, and understanding what are likely to be impacts all limit the immediate use of the SA and SGN as a simple tool. That they contradict each other in some fundamental aspects is hardly reassuring.

Reptiles

Three sources (Table A in the PSD, the reptile SA and SGN) differ in the recommended timing for detection. Table A generalises April to mid-October, the SA is more nuanced with April to May and September. Neither notes the use of a March visit to check on emergence from hibernacula, which is given in the standard methods guidance (Gent and Gibson 2003).

The SA requires desk survey data to help choose areas. The SGN omits this. The levels of survey detail required vary between the two guides. The SA clearly states its objective is to establish the population's size, not just presence or absence, which is important if impacts and mitigation are to be determined later in the process. In spite of that, the SA's limited coverage of timings, locations and limitations means that it is hard to determine the threshold for data quality. This is equally vague in the SGN.

The list of impact issues is clear in the SA but missing from the SGN, with the result that the SA is better in its explanation of mitigation. In spite of that, there are no post-development monitoring protocols in the SA, but there are some (without duration or scales) in the SGN. For those LPAs needing clarification, the SGN reference list is divided into topics. There are no references in the SA.

Overall, the clearest guidance comes from the standard reference Gent and Gibson (2003), not the SA and SGN, and an LPA would best be referred directly to the clear standard guidance in the first instance.

Bats

The SA and SGN draw on the same single standard reference: the Bat Conservation Trust (BCT) bat surveys for professional ecologists (Collins 2016). On first reading, Collins is daunting and heavily nuanced.

Neither the SA nor the SGN uses the same survey periods, while the SGN ignores the desk survey element of the PEA expected by both Collins and CIEEM (2019a). The range of methods covered in the SA is skewed towards visual detection or bat detectors. The SGN totally ignores the use of eDNA for bat droppings, which is an increasingly common method and cited in Collins (2016).

The SA provides a wider, more realistic, list of methods than SGN, but omits the licensing requirement for many of these. Any LPA reader would have to refer to Collins (2016) to fully understand survey needs and associated limitations. There are few references in SA, while the wide-ranging set in the SGN is poorly differentiated.

From the LPA standpoint, the two sets of guidance have different focuses and levels of explanation. To be certain of what was, or was not, done appropriately, the LPA would have to plough through Collins (2016) to begin to fully understand the inadequacy of the many planning applications that claim no impacts on bats.

SA and SGN: the conclusions

Given the highly variable set of ecological skills reported by ALGE (2013, 2020, 2022; Oxford 2012; Snell and Oxford 2022), and others, it would be expected that guidance from NE (SA) and a consortium including local planners (SGN) would be pitched at a level to meet the needs and expectations of LPAs. The stated aim of the SA that using the guidance alone would be enough to allow an LPA to make a decision is clearly unmet. This seems also to apply to the SGN, especially as the SGN omits full PEAs and impact evaluation.

3. Getting Better Data to Planners: Consultants, Data Quality and Constructing More Suitable Guidance

A lot of voluntary sector and statutory sector effort has gone into trying to improve the quality of ecological material provided to planners as part of planning applications. (e.g., BSI 2013; CIEEM 2018, 2019a,b; PBP 2021, Natural England 2022a,b).

As Chapters 1 and 2 indicated, the material provided to planners is far from perfect, and is directed at a corpus of planners with low levels of ecological competence and capacity (ALGE 2013, 2020; Snell and Oxford 2022; ENDS 2019; CIEEM 2018; DEFRA 2013). The result is that it would appear quite possible for poor quality material to routinely accompany planning applications, with a strong chance that problems with the data might not be readily understood (Reed 2021). That would be in spite of the best endeavours of a hard-pressed planning department.

Technically, as many of the ecologists that provide material for their clients are registered members of professional bodies, providing poor-quality data goes against their professional obligations and code of conduct (see Annex C BS 42020 (BSI 2013); CIEEM code of conduct (CIEEM 2022a).

With over 6,000 members, the Chartered Institute of Ecology and Environmental Management (CIEEM) is the leading Institute in the professional ecological field, and membership of CIEEM routinely crops up as a necessary criterion for employment by small and large consultancies alike. Membership is seen as a sign of basic value and repute by professionals and their employers.

Like other Institutes, CIEEM has a range of strictures that its members should meet. As part of its professional standards and codes, CIEEM routinely refers to the British Standard BS 42020: Biodiversity – Code of Practice for Planning and Development (BSI 2013) and the expectations that the standard lays upon members working as consultants when providing material for planning applications.

3.1 The British biodiversity standard BS 42020

Considering that the 2022 NE revisions of the SA, PPP and PSD all refer to BS 42020, it may help to note some of the key areas in BS 42020 and how these are meant to address, and preclude, potentially problematic data issues in planning submissions. With these in mind, it will then be possible to look at ways in which data gathering can be improved and the questions that need to be asked in guidance documentation, making for greater transparency on both sides of the planning discussion. If applied, and applied well, there should be few reasons to quibble over data provided in planning applications.

Much of the CIEEM Ecological Impact Checklist (CIEEM 2019a) is based on BS 42020, and because of this, and the problems of decision-making that stem from relying on poor guidance (on both ecologists' and planners' side of the desk), it will help to look at some of the elements in BS 42020 that a planner needs to be aware of, and why. For a full treatment, readers are referred to BS 42020 – preferably to be read alongside a planning application to see how and where, and if, it has been applied.

3.1.1 BS 42020: the basics

The central point of BS 42020 is that planning should be based on clear, well-documented data gathering and decision-making. In viewing the scope of the standard, BS 42020 states:

> This British Standard gives recommendations and provides guidance primarily for ensuring that actions and decisions taken at each stage of the planning process are informed by sufficient and appropriate ecological information.
>
> In particular, this British Standard provides recommendations and guidance to all professionals working in the planning and development sectors who might encounter biodiversity as an issue during the planning, design and development process on how to:
>
> a) meet obligations under codes of ethics or conduct when taking decisions or undertaking actions that could affect the natural environment; and
>
> b) adopt a professional, scientific and consistent approach to gathering, analysing, presenting and reviewing ecological information at key stages of the planning application process, or in evaluating the ecological implications of associated activities as part of consultation or other regulatory procedures.
>
> The processes recommended in this British Standard are applicable to the terrestrial, aquatic and marine environments.

That the standard opens with ethics for decision-making and the need for a consistent scientific approach is interesting. By definition, consultants should be acting professionally at all times, and that will determine the proper collection and assessment of data presented to planners. That in turn places emphasis on transparency at all stages.

In Section 4 – professional practice for biodiversity conservation – the standard sets out what this means, and why, a little more forcefully:

> 4.1 General
>
> 4.1.1. Professionals involved in both the preparation and determination of planning applications where biodiversity could be a material consideration should ensure that they have adequate access to appropriate ecological expertise in order to:
>
> a) establish whether any particular development proposal is likely to have a significant effect on biodiversity (See Annex A); and
>
> b) identify any measures necessary for compliance with all relevant statutory obligations and national and local planning policy.

This has carried extra weight since January 2022, as the new SA formally recognises all biodiversity as a material concern. The reference to appropriate ecological expertise is critical. For an ecologist, and specifically a member of CIEEM, this means that those carrying out the work on both sides should be demonstrably well qualified, something that ALGE established does not apply ecologically to many LPA planners. That gap is a fundamental problem when the 2022 SA has pushed the onus onto LPAs and away from consultants.

In 4.2.1 there is referral back to the professional and ethical stance of those involved. Aimed primarily at a developer's ecologists, it states:

> Where an individual is a member of an appropriate professional body they should act in accordance with their own code of professional conduct in all aspects of their work.

In 4.3.2 there is reference to technical competence: on both sides those involved in an application should have the competence and experience to carry out their tasks. This is something that many of those at the bottom layer of the consultancy market may have problems with (zu Ermgassen et al. 2021). Similarly, LPAs may also lack the necessary competence on their side of the desk, risking what could be a perfect storm of quality and confusion in both supply and planning determination.

Because of the limitations in data collected and offered in applications, and in understanding on the LPA side, there is a risk that emphasis and 'proofs' will often rest on the use of professional judgement. To an LPA planner, use of that term seems to offer some sort of reassurance (Reed 2019). Given that determination of planning should be based on a clear scientific audit trail, there is a risk that potentially unsupported professional judgement – which would be contrary to the CIEEM code of conduct – can be used in lieu of full data sets or a fully reasoned case (Reed 2019). In Sections 4.4.1 to 4.4.4 of the standard, the need for transparency is set out clearly, explaining why professional judgement is not a short cut, although it seems to be in many planning applications (Reed 2019):

> 4.4.1. Development proposals that are likely to affect biodiversity should be informed by expert advice. This should be based on objective professional

judgement informed by sound scientific method and evidence, and be clearly justified through documented reasoning.

4.4.2. In order to demonstrate sound professional judgement, the professional should:

a) gather all relevant information;
b) identify the issue(s);
c) identify practicable options for action that could be taken to address the issue(s);
d) make clear the weight to be attached to the issues and options considered; and
e) choose appropriate options and present these in a succinct and transparent manner, ensuring that the final decision or recommendation is clearly explained and can be justified.

This need for explanation and justification in 4.4.2.e is especially important where impact assessment is required, and mitigation is developed. That is confirmed in 4.4.3:

4.4.3. An explanation, with evidence, of the assessment and decision-making process and the reasons for a particular course of action or piece of advice should be clearly documented and made available where required and/or necessary.

NOTE: It is especially important to provide evidence of how professional judgement has been applied where work does not follow, in full or in part, the recommendations set out in national good practice guidelines (see 6.3.4, 6.3.5 and 6.3.6).

Section 6.2 of the standard sets out the issues of sufficiency and accuracy. These include use of wider areas than the immediate site, suitably long datasets and recent data sets. Many of these are issues that were not covered properly in the SA or SGN and are basic to fully comprehending potential impacts.

As many of the reports produced by consultants refer to standard methods or good practice, the standard looks in detail at the extent to which departures from good practice are not referred to or if the effects of these are not substantiated. Substantiation of claimed good practice is covered in 4.4.4. Section 6.3.8 puts things well:

6.3.8. To achieve full scientific disclosure (see 6.10), where the use of guidance is only relevant in part, is not followed, of if only parts of it are followed:

a) this should be fully justified in accordance with Clause 4; and
b) both the benefits and limitations (see 6.7) arising from any partial use or departure from good practice should be reported in full.

NOTE To claim compliance with good practice, and then not to disclose any omission or departures from such good practice, might be interpreted as a misrepresentation of the facts and could be in breach of an individual's code of professional conduct (see Clause 4).

6.3.9. Ecological reports should describe the methods of study and analysis actually used, rather than describing only what is published in good practice guidelines.

NOTE Describing published good practice guidelines, rather than the actual methods used for the survey, could be considered misrepresenting the facts and in breach of an individual's code of professional ethics or conduct.

6.3.8 is simple and clear, as there may well be a difference between what was said – or apparently followed – and what was done. It is this omission of the limitations section that can be the undoing of many claims. The requirement for clear limitations is absent from almost all of the 2022 SA guidance.

The issue of limitations is a vexed one, and is rarely addressed in most planning applications – yet it is also an important (but often missing) part of SA and SGN guidance, as many methods have clear limitations that affect their results and viability. These will be critical to determining data suitability and potential impacts. In Section 6.7, Identifying Limitations, the standard notes:

6.7.1. To reduce uncertainty, and to achieve full scientific disclosure, those undertaking surveys and preparing ecological advice and reports should identify all relevant limitations relating to:

a) the methods used, including:
   1) personal competence, i.e., qualifications, training, skills, understanding, experience;
   2) inadequate resources (equipment and/or personnel);
   3) inadequate time spent surveying;
   4) inadequate data (e.g., arising from incomplete or inappropriate surveys) giving rise to lack of statistical robustness and higher uncertainties;
   5) use of old and out of date data;
   6) timing or seasonal constraints and suboptimal survey periods; and
   7) partial use of and/or departure from good practice guidelines; and
b) site conditions and other factors, including:
   1) adverse weather conditions;
   2) restricted access to a site or part of a site;
   3) unrealistic deadlines; and
   4) unproven or untested measures for mitigation and compensation.

6.7.2. Any limitations associated with work should be stated, with an explanation of their significance and any attempt made to overcome them. The consequences of any such limitations on the soundness of the main findings and recommendations in the report should be made clear.

NOTE Failure to report limitations might be considered as misrepresenting the facts, and/or making erroneous assertions, exaggerated or unwarranted statements and therefore in breach of an individual's code of conduct (see Clause 6).

That last caveat note is critical, and 6.7.2 is rarely applied (Reed 2021). In their absence, claims are just that: claims, and personal opinions (Thompson et al. 2016). More than that, they could compromise an errant ecologist's professional standing.

These matters were earlier reinforced in section 6.2, which had noted:

6.2 Adequacy of ecological information

6.2.1. All ecological information should be prepared and presented so that it is fit to inform the decision-making process (see 8.1). As such, all ecological information should be:

a) appropriate for the purpose intended and obtained using appropriate scientific methods of ecological investigation and study (see 6.10);
b) sufficient, i.e., in terms of:
   1) scope of study;
   2) habitats likely to be affected;
   3) species likely to be affected;
   4) ecological processes upon which habitats and species and ecosystem function are dependent;
   5) coverage of a sufficiently wide area of study commensurate with the requirement of the species or feature of interest, including connected systems (e.g., downstream);
   6) undertaken over a sufficient period of time and at an appropriate time of the year to reveal sufficient details of population or habitat characteristics (see 6.4);
   7) being sufficiently up to date (e.g., not normally more than two/three years old, or as stipulated in good practice guidance); and
   8) identification of risks e.g., spread of pathogens or invasive non-native species.

NOTE The shelf life of any given survey depends on the type of survey undertaken and whether environmental conditions within the study area were "normal" or unusual at the time undertaken (e.g., extreme weather), or are likely to have changed or remained the same. The greater the recent change, the greater the need for up-to-date information.

Just in case an applicant's ecologist missed 6.7, Section 6.10, 'Full disclosure of scientific method', requires:

6.10.1. The evidence underpinning all ecological advice and reports should be robust and obtained using reproducible scientific methods that allow the reliability of data to be verified.

NOTE 1 Such practice is called 'full disclosure'. Scientific experiment or study needs to be capable of being accurately reproduced or replicated by someone else working independently. This enables careful scrutiny (see 8.2) by other ecologists and professionals, giving them the opportunity to verify results and to analyse and interpret them independently. This is one of the main principles of scientific method.

NOTE 2 There are many reasons, and increasing demands, for full disclosure of the underlying data used to support ecological opinions. This is even more important where there is uncertainty or scepticism by the public, third parties or other professionals over the claims sometimes made in ecological reports. Full disclosure helps reduce uncertainly and scepticism (see 6.6 and 8.2).

6.10.2. Ecological judgment and advice should be based on sound scientific principles and be as objective as possible to avoid biased, unwarranted or exaggerated interpretation of the results presented. To achieve this, ecological results should be based on the:

a) identification of a set of relevant scientific (ecological) questions and the design/selection of appropriate methods to answer these;
b) appropriate implementation of the selected methods;
c) objective analysis of data gathered, ensuring that all information is fit for purpose; and
d) impartial interpretation and presentation of results that enable valid conclusions to be drawn and justifiable recommendations to be made.

NOTE 1 These elements of scientific method are also incorporated into the CIEEM Professional Competency Framework.

Item 6.10.2 (a) is central to any use of SA and applies equally to consultants or the advice that an LPA planner might get from internal or contracted ecological advisers: know what you are trying to achieve and be sure that the methods are suitable, and the data sought are fit for purpose. And, of course, make sure that the interpretation is based on those data and is credible. Given the problems with the January 2022 SA guidance, it is unclear how easily an LPA planner could determine this.

All of these requirements appear again in Section 8, 'Decision-making', especially in 8.1, 'Making decisions based on adequate information'. The basic point is quite simple: if the surveys, data and reliability are at question, then you have not got adequate information. In which case, you either start again, or remove the application.

Section 8 relies upon the adequacy of staff at LPAs to determine whether what they have in front of them is correct. If hoping to rely on the 2022 SA for this, then there is almost no way that 8.1 could be decided on the basis of those 2022 tools.

8.1 Making decisions based upon adequate information

The decision-maker should undertake a thorough analysis of the applicant's ecological report as part of the wider determination of the application.

In reaching a decision, the decision-maker should take the following into account:

a) the soundness and technical content of ecological information, to ensure:
   1) the proposal[s] are based on adequate (see 6.2) and up-to-date ecological field data that substantiate clearly the conclusions reached and recommendations made;
   2) ecological methods are, where available, in accordance with good practice guidance (see 6.3.6); and
   3) departures from any good practice are made clear, are valid and can be justified (see 4.4, 6.3.6 and 6.3.7)
b) Whether biodiversity is likely to be affected and whether all potential impacts are described adequately, for example in relation to:
   1) location and extent;
   2) timing and frequency;
   3) duration;
   4) scale or magnitude;
   5) reversibility/recoverability/resilience;
   6) in-combination/cumulative effects; and
   7) likelihood/degree of certainty associated with predicted effects.
c) Whether effects are significant and, if so, capable of being mitigated.
d) Whether the mitigation hierarchy has been applied (see 5.2).
e) Whether it has been adequately demonstrated that the proposals will deliver the stated outcomes if consent is granted, with particular regard to:
   1) likely effectiveness, e.g., proposed ecological measures are appropriate to the case and technically feasible, and if implemented, likely to achieve desired outcomes; and
   2) certainty over deliverability, e.g., there is evidence of commitment and adequate mechanisms to secure sufficient land and resources to implement necessary measures.
f) Whether the measures are capable of being secured through appropriate planning conditions and/or obligations (see 9.2, 9.4 and Annex D) and/or likely to be permitted through another consent regime, e.g., licences for European protected species (See 9.5, Annex D and Annex E).
g) Whether the proposals are compliant with statutory obligations and policy considerations (See Annex B).
h) Whether there is a clear indication of likely significant losses and gains for biodiversity.
i) Whether any material considerations have been identified that might require changes to the application.

All of this depends on properly collected data, using properly phrased questions and there being proper guidance available to the planners. If this is not the case, then nothing can be done.

In the next chapter, mindful of the very clear requirements of CIEEM-registered consultants set out in BS 42020, we will look at the 2022 SA afresh and go back to the initial questions that anyone looking at a site should be asking, and how to address these in practice.

4. The New 2022 Standing Advice: Turning Around and Moving Forwards

The 2022 revision of the NE documents (Natural England 2022, 2022a,b) transfers the 'burden' (Gove 2018) from the shoulders of developers onto the shoulders of the LPA. For the applicant – newly titled the developer – the need to show impacts, a term no longer used, is less categorical. With the 2022 revisions, the role of the SA has also gone from being used to 'decide', to the less categorical 'consider'. How that fits with biodiversity being a material consideration is uncertain.

In the following sections I will consider how an LPA can try and address its duties based on a further improvement to the SA approach.

4.1 Establishing the facts

In order to understand the potential effects or impacts of a development upon an area, both the developer and the LPA need to have a set of guidance that they understand and use. This must produce material that is clear and uses methods that are stated, applied and where limitations are recognised. This means that the developer's agents (consultants/in-house ecologists) must apply the tenets of BS 42020, and in a way that the LPA can understand what is being said and meant.

It is critical that the LPA has a set of guidance that allows it to double-check what the developer is claiming. That was not provided by the pre-2022 SA and is most certainly not provided by the January 2022 SA. Nor did the SGN fully provide it. This requires a fourth way.

A simple approach is set out in the following sections. With such an approach in place, there is a reasonable chance that both sides will be able to operate with the same set of facts and assess impacts and the suitability of a proposed development, thus providing a basis for discussing biodiversity net gain (BNG) too. Without it, any meaningful discussions of proposals or BNG will be largely illusory.

The first step is to recap what is known, and what needs to be known, so that developers and LPAs can work to a common standard.

We know that:

- planning applications should be accompanied by biodiversity data that provide an unequivocal assessment of the current resource in/around the site;
- these data should normally be less than two years old;
- data collection should be based on a set of stated needs, normally given by planning authorities in a validation checklist;
- all ecological data should be collected in line with the precepts of BS 42020;
- data collection should be undertaken using stated methods and how these methods were applied should be made clear; it is not enough to say method X was used. It has to be shown that it was;
- any issues that mean limitations occur in the application of methods need to be stated and their effects on the dataset evaluated. Stating that limitations do not affect the data is an act of faith, not fact, and it needs demonstrating why they do not apply;
- site data do not exist in isolation. Effects are normally cumulative, and impacts need to address this;
- the spatial context of a site is critical to understanding its effects on the surroundings and vice versa;
- protestations of no impact need to be backed up by data to demonstrate this is well founded;
- being able to demonstrate a potential level of impact is the basis for understanding what needs to be mitigated or compensated for;
- the level of reliability in habitat surveys that underpin biodiversity net gain is not high (zu Ermgassen et al. 2021); the degree of non-repeatability in habitat surveys that are to be used for biodiversity net gain is in excess of the ≥10% gain being sought by biodiversity net gain;
- biodiversity net gain focuses on habitat rather than species and is still work in progress at the time of writing (September 2023);
- data need to be collected and provided in such a way that they can be added together to make a coherent larger picture; something that seems a challenge to many ecologists (zu Ermgassen et al. 2021);
- planners need to have the competence and capacity to understand what is provided to them, and to comprehend whether what is being claimed is fact or illusion;
- the majority of planners are not well-versed enough to understand the nuanced needs and claims that they encounter, and that is by their own admission;
- from 2023 onwards, the demands on planners from planning applications will be more, rather than less, as the 'burden' is switched from developers to planners;
- if demands are to be increased, then LPA competence and capacity will need to be increased in turn;

- BNG consultations expect a strengthened LPA body to address BNG issues and suggest that in order to do this more resources would be available to LPAs. That is to be tested in the future;
- stating that biodiversity is a material planning consideration that is only required to be taken into account among other factors is to weaken the hand of LPAs;
- any further demands on a beleaguered LPA planner will require credible guidance tools to be available;
- in England, the January 2022 SA has reduced the already questionable level of quality in pre-2022 SA, weakening any basis for biodiversity delivery. The contrast with NatureScot guidance for protected species is striking;
- LPAs need better guidance than the January 2022 SA offers, or is currently available from the SGN;
- melding the best aspects of the SA and SGN seems a pragmatic way forward, as well as adding missing elements. Being informed by NatureScot guidance offers a positive way forward.

The following sections look at what an LPA needs, and seeks to fill the gaps, starting with what an LPA needs to know and why.

4.2 What does an LPA need to have?

At its simplest, an LPA planner needs to have:

- Competence and capacity to address data needs and planning submissions with confidence. That is something a guide cannot fix readily. Providing a set of tools that helps build competence will go a long way to improving the situation.
- A set of tools that will allow a planner to check through what was provided, and perhaps should have been provided, and to help check what should be available to the planner in a planning submission.

4.3 What does the January 2022 SA expect of an LPA planner?

Under the January 2022 SA, planners are now the new polymaths who are expected to have, or have access to:

- skills on survey types across all groups and species, especially protected species listed in the SA;
- be able to determine adequacy of data records and methods accompanying a planning application;
- be able determine the meaning in patchy historical records, mostly derived from the National Biodiversity Network (NBN);
- assess habitat suitability in and around a site;

- capacity to assess surveyor capacity/competence;
- assess the likelihood of effects on species/habitats, rather than the applicant (now called the developer), and how the developer should achieve net gain;
- competence in determining operation of the mitigation hierarchy and forward planning;
- expertise in monitoring and long-term change;
- the time to search the NBN and other data sources for records in and around the site and to know what they might mean biologically, context wise – or not.

As observed earlier, before 2022 those nine points were largely the preserve of the applicant but are now part of the LPA planner's burden.

In 2023, and most likely beyond, expecting LPA planners to meet more than half of these is unlikely. The 2022 review of LPA capacity to deal with Biodiversity Net Gain (Snell and Oxford 2022) confirms the inability of LPA staff to address BNG data due to competence and capacity issues, something noted by Robertson (2021) as well.

To begin to address any of the nine issues identified above requires a working version of the SGN and SA that provides the basic details required by planners, one that acts as a counterweight to claims in an application. It will also require rephrasing some of the burden in the SA, so that the PPP and PSD recognise what is the work of the developer's staff and what is credible/reasonable for the LPA to do.

Table 4.1 summarises the key features of the 2022 SA, and looks at what the SA does and does not provide. The SA does provide reference to legal protection, but the first real basic problem is that it does not provide viable survey guidance. This means that, if using the SA as a stand-alone set of guidance, the LPA planner would not know what was needed in an application, nor what was missing. To help fill this gap, an LPA planner might reasonably look for the standard survey reference for the species. In almost all cases this is guidance that is not referenced in the revised SA guidance, leaving the planner badly exposed by the 2022 SA.

Under the January 2022 SA, planners are expected to know whether there are suitable habitats in the area, in which case, it might have helped to list these for each species. Some SA entries hint at them, but most ignore them.

Almost all SA entries require LPAs to use the NBN for presence or absence of species in the immediate area of the site. The SA documents do note that most of the data in the NBN are problematic: absence of records is not record of absence. As a result, it is hard to interpret many of the searches from NBN data.

The SA entries give no guidance for suitable distances to search for in the NBN, and refer to the suitability of data age for use. Interestingly, the standard PEA guidance (CIEEM and NE) still refers to data searches from Local Record Centres (LRCs). From personal experience, NBN data do not necessarily match LRC data. There is also the uncertain ground of whether NBN data can be used for commercial purposes; it cannot according to NBN (https://nbn.org.uk/). The issue here is whether the LPA is in effect acting in a commercial sense. As it is viewing NBN data to validate a commercial proposition – a planning application – it might be argued that it is commercial assessment by proxy.

When considering potential effects of a development (in the new SA there are no impacts), and any use of elements of the mitigation hierarchy, all of the entries are so generic that they cannot be pinned down for practical use. The absence of numeric, temporal or qualified material leaves a reader wanting more: unable to use what is provided. Mitigation options are so general as to be almost tick lists and have no caveats on scale, duration or any approximation to biological reality. The nadir is reached with the biodiversity net gain options: green roofs, street trees and SUDS being offered across the board, with no thought to possible ecological applicability.[1] Can these really apply to Freshwater Pearl Mussel, Otter and Hazel Dormouse?

The result of looking at Table 4.1 is that LPA planners cannot credibly use the 2022 SA to undertake their role as set out in the SA.

4.4 Moving forwards from the 2022 SA in 11 steps

If LPA planners cannot rely on the old SA, the SGN or the quite clearly inadequate 2022 SAs, then how would they be expected to deal with understanding what they are routinely presented with in a planning application?

Ideally, the biodiversity element of a fully credible planning application would have followed all of BS 42020 and be clear, concise, reasoned and verifiable, list its limitations and impacts, and have clearly applied the mitigation hierarchy. In the verifiable likelihood that there were still negative impacts associated with the development after demonstrable efforts had been made, and to address these a credible basis for biodiversity net gain had been presented followed by a fully costed and timetabled programme of monitoring and post development management, then the role of the LPA would be straightforward. As this so rarely happens in practice, a planner needs to be wary and able to refer to a simple set of short-cutting steps with which to appraise an application. Where probity is at a premium, planners need a way of smelling it out. That means that poor applications need to be detectable too.

In the following section, I look at what is needed, but missing, from the old and new SA, and the steps that NE states should be followed in general, plus the questions to ask (in line with BS 42020) to be certain of the level of detail needed in a credible application.

Step 1: who starts the process?

Who initiates the need to use the SA guidance? The need for a planner to consider protected species advice does not come from out of the blue. Rather, it comes as part of an application. The confusion between the 2022 PSD – which expects LPAs to go out to would-be developers in advance of an application – and the 2022 revision of

---

[1] Sustainable drainage systems (SUDS) are drainage solutions that provide an alternative to the direct channelling of surface water through networks of pipes and sewers to nearby watercourses.

Table 4.1 The key features of the suite of January 2022 SA guidance

Category	Badger	Bats	Birds	Fish	Pearl Mussel	GC Newt	Hazel Dormouse	Natterjack Toad	Otter	Protected plants, fungi and lichens	Reptiles	Water Vole	White-clawed Crayfish	Ancient wood/tree/ veteran tree
SA gives outline of protection/offences	*	*	*	*	*	*	*	*	*	*	*	*	*	*
SA provides guidance on survey methods	–	–	–	*	–	–1	*2	*2	–	–	*2	*	*2	–
LPA required to search NBN for distributional data	*	*	*	*	*	*	*	*	*	*	*	*	*	–
LPA required to know if suitable habitat types present for possible survey request to developer	*	*	*	*	*	*	*	*	*	–	*	*	*	–
LPA required to assess competencies of applicant staff using CIEEM criteria	*	*	*	*	n/a	*	*	*	*	*	*	*	*	–
LPA to use CIEEM guidance to determine validity of data age	*	*	*	*	*	*	*	*	*	–	*	*	*	–
SA lists limited range of possible avoidance methods	*	*	*	*	*	–	*	*	*	*	*	*	*	*
If avoidance impossible, developer to show mitigation options	*	*	*	*	*	–	*	*	*	*	*	*	*	*

SA lists limited range of mitigation measures	*	*	*	*	*	*	*	*	*
SA mentions licence possibility and link provided	*	–	*	*	*	*	*	*	–
SA gives limited generic enhancement list	*	*	*	*	*	*	*	*	–
SA gives generic biodiversity net gain list	*	Unsuitable	Unsuitable	*	Unsuitable	*	Unsuitable	Unsuitable	Unsuitable
SA gives limited notes on site management and monitoring	*	*	*	*	*	*	*	*	–
SA provides reference to external guidance	–	*	–	–	–	–	*	–	*

1 = district licensing scheme preferred by NE; 2 = survey methods inadequate for valid use and reliance unsuitable: green roofs, street trees and SUDS unlikely to benefit. *An entry under the category title; – entry missing under the category title. SA = standing advice; LPA = local planning authority; NBN = National Biodiversity Network; CIEEM = Chartered Institute of Ecological and Environmental Management

the PPP – which notes that developers go to LPAs as part of the planning process to which the LPA may respond by asking for a PEA and a range of details (including for those species covered by SA) – makes it hard to be categorical.

For simplicity, and in line with many VCs, it will be assumed that a would-be developer may initially approach an LPA to discuss a proposal in outline (as in Scotland), and to seek guidance, without prejudice to either side. That may subsequently be formalised or clarified with reference to the authority's VC guidance, if available. The CIEEM (2019a) process checklist covers part of this, but cuts across both of the more recent PPP and PSD. Under the CIEEM (2019a) approach, protected species should be picked up at the earliest stages of a developer's consideration of what needs to be covered in an application. That is echoed in the PPP.

The key question in step 1 is who determines the start of the process? Is it the developer or the LPA? The PSD infers it is the LPA. Illogical, and as contradictory to the PPP as this may seem, it chimes with the SA which states that the LPA should ask for a survey if its (the LPA's) searches of NBN data or discovery of suitable habitats in and around a site (how is left unstated) uncover a protected species covered by an SA note.

So, step 1 seems to have two answers: the developer approaches the LPA or vice versa. We will assume that it is the developer that approaches the LPA, and that the developer may mention the possibility of a protected species in the vicinity. This may well match the knowledge base of the LPA.

Resolution: it will be assumed that the developer will approach the LPA. It is normally required that the LPA will have a valid VC to guide considerations.

Step 2: Who initiates the search for records and suitable habitats?

Having somehow started the process, the next step according to both the PSD and the SA – but not the PPP, is for the LPA to decide if there is need to ask for a survey. The bases for this are either:

- Historical or distribution records suggest that [Badgers] are active in the area – you can search the National Biodiversity Network Atlas by species and location.
- There is suitable habitat for [sett building or foraging].

Both would need caveats on distance from the core of the site – especially for wide ranging species – and recency (Natural England 2022b).

How, or why, the LPA would be responsible for ascertaining and determining presence of suitable habitat, and at what scale, is left hanging. Individual SA entries qualify the base for surveys but not normally in a way that provides quantitative data.

Resolution: the developer provides the record search and habitat details as part of the VC process. Outline details for each species covered by a SA are given in the tables in Chapter 5, based on individual species survey standards and reports. The default distance for a data search would be 2 km.

Step 3: who is qualified and licenced to do the work?

In this step, there is a further case of NE documents reversing roles. In the PPP, the next step, assuming that a protected species occurs on or near the development site – and there is no guidance as to distance to be used – is to undertake an assessment of whether the planned development would affect the species. If so, a survey is needed to confirm this. In the PSD and SA, a survey would be needed. In which case, they call for a suitably qualified and experienced ecologist. CIEEM competencies are cited. As many protected species also require ecologists to hold licences to disturb the species as part of the survey process, competencies and experience may of themselves not be enough. The PPP does not mention this.

Resolution: where ecologists need to hold survey licences in order to disturb species as part of the process, this is noted in the tables in Chapter 5. The LPA should not be responsible for assessing the competencies of an applicant de novo. That is for the developer to do as part of the application support documents. Licence status of ecologists should also be provided.

Step 4: Assess the effect of the proposed development on the species

Before the scale of mitigation can be estimated, the proposal needs to show the extent to which the species might be affected. In the PPP, this is about undertaking surveys. In the majority of the SA entries, there is a total lack of survey guidance. What exists is so generic that critical details that make or break the reliability of the surveys in many applications are missing. That is contrary to the expectations of BS 42020. In addition, in the SA, there is an almost total failure to link to standard survey documents that LPAs will need to be aware of when evaluating applications.

Understanding the scale of potential effects is critical to understand the possible licensing issues that will come later in the sequence. As part of the licencing process (step 10), there is a requirement for a method statement. This requires a clear description of how the proposed development might impact the species – in the case of Badgers, this would mainly be the sett but would also include foraging habitat – and how this would be mitigated. This also requires details of survey methods and results and the basis on which territorial ranges are determined. None of this is covered in enough detail in the new SA.

Resolution: details of survey methods, either missing or understated by previous SA and SGN, are given for protected species in the tables in Chapter 5. These form the basis for appraising the quality or reliability and probity of data submitted to an LPA. These are needed for step 10, which cannot be relied on if using SA information alone.

Step 5: Are the data going to be as robust as claimed?

Before involving the use of the mitigation hierarchy (avoidance, mitigation and compensation), any review needs to be sure that the data are robust enough to rely on and can be used to support claims in the application. As stated by BS 42020, this means that an application will need to include honest appraisals of limits, and other factors, that may affect the reliability of impact claims. This will need to include site-specific effects, effects external to the site and cumulative impacts. It will also require a detailed summary of the methods undertaken, not just noting that standard methods were applied.

Resolution: consider factors that may be limitations in the initial review of data collection and data quality and assess against survey methods in the tables in Chapter 5.

Step 6: Does the application really use avoidance in the mitigation hierarchy sequence?

The PPP requires avoidance. The PSD is so generic that there is no guidance on how to evaluate what this might mean, spatially and temporally, and how it would be determined. Before attempting to use the mitigation hierarchy, a thorough and reliable set of survey data is needed. Without it, unsupported professional opinion is not a viable alternative, and mitigation cannot be started. Step 5 may need revisiting.

Resolution: establish a spatially viable baseline with suitable data collection in order to establish which areas used might potentially be affected; see tables in Chapter 5.

Step 7: What is meant by suitable mitigation measures?

Mitigation promises to reduce development effects significantly. Many mitigation methods are generic, and need referring to *Conservation Evidence* or similar literature for their efficacy before use, and need supportive references before promotion (Hunter et al. 2021). Picking out an apparent mitigation option from a generic list in the SA, such as bird or bat boxes, is normally devoid of ecological or spatial context that would help decide suitability; these options need to be linked to other wider measures too. It should be noted that not that many species will actually opt for boxes, and placement is key. The extent of use by bats is also questionable (Lintott and Mathews 2018).

The 2022 PSD requests that mitigation measures are: adequate, effective, established before development starts, reliable, measurable and secured. Both the 2022 PPP and PSD link to a mitigation plan (HMG 2022), but this is also a generic overview of how to approach things and omits any informed examples for use. The method statement in step 10 is needed here too.

There is one major problem that needs to be addressed. It is assumed that proposed mitigation measures being cited are valid. Hunter et al. (2021) looked at a sample of 50 UK housing applications with 446 associated mitigation measures.

A total of 65 different mitigation measures were put forward for eight taxa. Most (56%) were justified by citing published guidance. Hunter et al. (2021) looked at the literature behind the proposed ecological mitigation and compensation measures (EMC). Less than 10% had empirical data. In addition, citation network analysis also identified circular referencing across bat, amphibian and reptile EMC guidance. Looking at *Conservation Evidence* synopses showed that over half of measures recommended in ecological reports had not been empirically evaluated; only 13 were assessed as beneficial.

Put simply, absence of evidence on the effectiveness of EMC measures makes the basis of glib assertions for compensation just that bit more problematic. It also underlines the need to avoid impacts from the outset, rather than assume that compensation will solve all ills. Biodiversity net gain is thus on shakier ground than the government might have us believe if both the survey data and the compensation bases are potentially spurious.

Hunter et al. (2021) was not the only group doubting the factual basis of BNG. The BTO looked at the data available for BNG in an urban setting (Plummer 2022). As they put it:

> In England, most new infrastructure developments will soon be legally required to have a positive impact on biodiversity (providing 'biodiversity net gain'), and similar strategies are in place across the UK. But the robust evidence base necessary to support the delivery of these biodiversity-rich urban areas is lacking.

That is hardly a ringing endorsement of BNG.

Resolution: consider the ecological value that will be gained, and show how positive effects will be achieved and over what timescale. Double check that mitigation measures cited are supported in Hunter et al. (2021). Look in the tables in Chapter 5 for some qualified species-specific options.

Step 8: Consider how to compensate for negative effects

The PPP describes compensation 'As a last resort you should compensate for impacts', but puts forward no details, while the PSD puts it to the LPA thus:

> If avoidance or mitigation measures are not possible, as a last resort you should agree compensation measures with the developer and put these in place as part of the planning permission.

Last resorts are not the place to be. Compensation depends on knowing what is there and what will be affected and what is actually being compensated for. This requires all of the earlier steps to be done, and done well, before this can be estimated. For LPAs, being sure that estimates for what is to be compensated are credible requires better data than are usually provided. As noted above (Hunter et al. 2021), compensation is frequently more faith-based than fact-based.

The PSD requires no net loss, increased connectivity, like-for-like habitat replacements adjacent to existing populations or alternative locations. The PSD expects these areas to be in place to provide havens before development starts. Section 106 agreements offer the chance of delivering and monitoring long-term changes, but need linking to realistic numbers in a plan, numbers that may be problematic to collect from the outset.[2]

---

[2] Section 106 (S106) Agreements are legal agreements between local authorities and developers; these are linked to planning permissions and can also be known as planning obligations. Section 106 Agreements are drafted when it is considered that a development will have significant impacts on the local area that cannot be moderated by means of conditions attached to a planning decision.

Resolution: LPAs need to check that data provided are consistent with tables in Chapter 5 and that the claims in an application have some validity. Otherwise, compensation is a chimera.

Step 9: Consider enhancement

Enhancing habitats within the development site is advocated by the PSD, using generic tools of little relevance to species covered by the SA. While the PSD does not mention the concept of net gain, the PPP, launched on the same date in 2022, does, but the link to the NPPF (HMG 2021) provides little guidance. Instead, the PPP lists actions that were covered under *compensation* in the PSD; another case of misunderstanding between the two sets of guidance.

As the NPPF requires ≥10% net gain, it seems odd that the two documents are unable to decide where, what or how this should be achieved. Again, this gain must relate to good-quality data with a linked management regime and losses and gains shown. Enhancement has winners and losers, and these need to be identified and spelt out. As a form of compensation, enhancement needs also to be demonstrably valid and viable.

Zu Ermgassen et al. (2020, 2021) noted that much of the 'benefit' to net gain was from transforming grassland within a development into scrub or woodland, rather than adding extended areas of habitat as the result of a development.

Resolution: Confirm what biodiversity resources are on the site, and how these might be enhanced without disbenefit for protected species. For SA, it will take careful examination of ecological communities and spatial scales, and valid, fact-based options.

Step 10: Is a licence needed?

Because many of the species that might be affected by a proposed development are protected, and that clearly applies to protected species covered by the SA/PSD, if a development wants to proceed, then there will normally need to be an application for a licence to allow this. The PSD identifies that licences and planning permissions are different. As the PSD puts it:

> Licences are subject to separate processes and specific policy and legal tests.
>
> You should tell the developer if they're likely to need a protected species licence from Natural England or DEFRA to allow activities that would otherwise be illegal.
>
> You must be satisfied that if a licence is needed it's likely to be granted by Natural England or DEFRA before you give planning permission.

How that is done is left unstated. The PPP issues similar guidance and cross-refers to the PSD and the paragraph shown above, so that the clarity hoped for is lacking. It all hinges on material not available in the SA.

European protected species (EPS) still have cover post-Brexit. In order to obtain an EPS licence, much of the basis for application is based on survey and other material no longer provided for in the 2022 SA.

Resolution: For an applicant using the PPP guidance, and for a LPA using the PSD, the strength of the material provided in a method statement and other supporting material cannot be relied upon if the sole source is the 2022 SA. Having access to more detailed material, of the sort found in the NS species guidance, would help immensely, and can be found in the tables below. The LPA has to decide if a licence is needed before permission can be given; that is outside of the locus of the SA.

Step 11: An LPA makes a planning decision

By the time an applicant has got to the licensing stage, its job is pretty well done – for the first time at least, according to the PPP which is now finished. In the PSD, the last, and final, stage 5 is listed as follows:

5. Make a decision about a planning application

If the proposal is likely to affect a protected species, you can grant planning permission where:

- a qualified ecologist has carried out an appropriate survey (where needed) at the correct time of year
- there is enough information to assess the impact on protected species
- all appropriate avoidance and mitigation measures have been incorporated into the development and appropriately secured
- a protected species licence is needed it is likely to be granted by Natural England or DEFRA
- any compensation measures are acceptable and can be put in place
- monitoring and review plans are in place, where appropriate
- all wider planning considerations are met.

This is effectively a game of snakes and ladders, as the very material required for almost all of these elements is missing from the 2022 SA (Table 4.2). How these elements are to be done, and found acceptable, is uncertain from the PSD or the SA. In a hope to bolster a confused LPA, the PSD states: 'The decision-making checklist (PDF, 162 KB, 4 pages) can help support planning decisions.'

Can it? Does it? Most of the checklist is reliant on the material missing from the SA. For a poorly qualified LPA planner, or in an LPA lacking the necessary capacity and/ or competence, the list can help, but only in so far as the right material is available to planners and can be recognised as such. That is in question.

The 2022 ALGE report on LPAs and BNG (Snell and Oxford 2022) indicates that the position at LPAs identified a decade before (ALGE 2013) has not changed for the good. A total of 5% of the 337 staff in 192 LPAs responding to the 2021 survey on biodiversity capacity and competence said their current resources were adequate to scrutinise applications that might affect biodiversity. Fewer than 10% said their current resources and expertise would be adequate to deliver BNG. That means that there is a clear need for a guide to help LPAs address issues from scratch, especially as Snell and Oxford (2022) stated that the ALGE (2013) finding that 90% of LPA

Table 4.2 Issues with using the NE 2022 SA guidance

Category	Badger	Bats	Birds	Fish	Pearl Mussel	GC Newt	Hazel Dormouse	Natterjack Toad	Otter	Protected plants, fungi and lichens	Reptiles	Water Vole	White-clawed Crayfish	Ancient wood/tree /veteran tree
SA gives outline of protection/offences	*	*	*	*	*	*	*	*	*	*	*	*	*	*
SA provides guidance on survey methods	–	–	–	*	–	–	*2	*2	–	–	*2	*	*2	–
LPA required to search NBN for distributional data	*	*	*	*	*	*	*	*	*	*	*	*	*	–
LPA required to know if suitable habitat types present for possible survey request to developer	*	*	*	*	*	*	*	*	*	–	*	*	*	–
LPA required to assess competencies of applicant staff using CIEEM criteria	*	*	*	*	n/a	*	*	*	*	*	*	*	*	–
LPA to use CIEEM guidance to determine validity of data age	*	*	*	*	*	*	*	*	*	–	*	*	*	–
SA lists limited range of possible avoidance methods	*	*	*	*	*	–	*	*	*	*	*	*	*	*
If avoidance impossible, developer to show mitigation options	*	*	*	*	*	–	*	*	*	*	*	*	*	*

SA lists limited range of mitigation measures	*	*	*	*	—	*	*	*	*
SA mentions Licence possibility and link provided	*	—	*	*	*	*	*	*	—
SA gives limited generic Enhancement list	*	*	*	*	*	*	*	*	—
SA gives generic biodiversity net gain list	*	*	Unsuitable	Unsuitable	*	Unsuitable	*	Unsuitable	Unsuitable
SA gives limited notes on site management and monitoring	*	*	*	*	*	*	*	*	—
SA provides reference to external guidance	—	*	—	—	*	—	*	—	*

1 = district licensing scheme preferred by NE; 2 = survey methods inadequate for valid use and reliance unsuitable: green roofs, street trees and SUDS unlikely to benefit. *An entry under the category title; – entry missing under the category title. SA = standing advice; LPA = local planning authority; NBN = National Biodiversity Network; CIEEM = Chartered Institute of Ecological and Environmental Management

planners lacked ecological qualifications, and had minimal biodiversity training, had not changed a decade later. For those LPAs without ecological resources, Snell and Oxford (2022) noted reliance on personal judgement and the NE SA. The latter is clearly untenable, and the former is unsafe due to competency issues.

Resolution: an LPA needs to check all of the methodological material available to it, using sources other than the 2022 SA, and also needs to be sure whether it asked for the right material from the outset. All of the material needs checking with proofs of adequacy – see the earlier discussions on limitations and field methods – and a clear idea of what needs to be conditioned, and why, and how this will be monitored effectively. Whether a poorly qualified LPA planner – given capacity and competence issues (Snell and Oxford 2022) – can do this is a basic problem. The following set of species-specific tables in Chapter 5 should help ease the LPA burden.

5. Surveys for Protected Species: What the LPA Might Have Ordered

If the guidance offered by the 2022 NE species SA is inadequate for purpose, and the material in the SGN is also of limited value, what is an LPA planner meant to do? It is clear that neither the 2022 PPP nor the 2022 PSD is robust and that they operate in conflict. On that basis, the following set of outline guidance is offered to help an LPA planner through the quicksand of data provision and evaluation. It is hoped that by applying the approach set out in the following tables, it will be possible for an LPA planner to make their way through the material that might come to them and compare it with what they should rightly expect. If it is inadequate, then, under the PPP and PSD, the application should be rejected until proper data are supplied. The challenge is to know – something that the SA fails to show.

The provision of suitable data, using clear methods is the bedrock of planning evaluation. A PEA will indicate some of the species/issues to be covered. Detailed surveys in an EcIA (phase 2) will often be needed, and the tables (Table 5.1 *et seq.*) in this chapter indicate the sorts of detail required (CIEEM 2017).

5.1 Headings: what to expect and why

Before getting to the missing details in the SA, it is important that the LPA goes through some of the early steps noted above in Chapter 4. These include informal discussions and pre-application advice.

Planning is a two-way process: an LPA has policies and procedures to apply, and applications should fit their structures at the local level and ultimately be consistent with national policy requirements according to the PPP and PSD. The PPP notes that it is commonplace for applicants to have a first discussion with LPA planners, and for both to begin to understand the aspirations of the other. For an applicant, this might well include accessing the authority's VC (Abrahams 2019). It would also set the tone of expectations and provide the LPA with an understanding of the capabilities of the would-be applicant.

Recording when, where and who was consulted, and if a VC was covered, is an important first step in the transparent audit process for both planner and applicant. Any other discussions with other statutory bodies or leading NGOs should also be noted.

5.1.1 PEA?

An important part of an application is the context: knowing where the site is, what is on it and what habitats, species and protected sites are in the area will inform discussions, and help in understanding potential immediate and cumulative impacts. Both the 2022 PPP and PSD mention the possible use of a PEA, while the 2022 SA does not. As most LPAs will require a PEA under their VCs, it seems odd that the 2022 SA guidance from NE ignores this completely.

From the LPA standpoint, a PEA or EcIA should be expected, and LPAs should refer to CIEEM (2017) for contents.

The key element of the PEA is the desk search: normally a commissioned search of data held by LRCs and usually a separate check of local bat groups – although these may now, as in Bedfordshire, increasingly be the same as bat group records are lodged with LRCs. In addition, some initial scoping may be done via the NBN – although there is explicit wording on the NBN site that it should not be used for commercial purposes. That appears not to preclude their use by LPAs for what are technically non-commercial purposes. This needs confirmation.

As well as the desk data, which was noted earlier as being varied in quality and reliability, a PEA should include biological data from a walkover survey. Such data have their own limitations (see BS 42020 (BSI 2013) and CIEEM 2017). These will be examined in the SA contents below. For many sites, a PEA will indicate the need for more detailed (phase 2) surveys to inform an EcIA (CIEEM 2017). Table 5.1 *et seq.* should be consulted for protected species.

5.1.2 Site name and geographic context

Every site has a name and sits within a functional biological context. This could be a plot in a floodplain or a block of moorland adjacent to lower-level damp meadows. Typically, species on a site form part of a wider continuum, connected to areas for feeding or breeding or as a population cycling in time and space. This effective zone of influence is important because impacts beyond the immediate area of a proposed development will potentially have significant effects on species occurring there in whole or part (Hounsome 2021). Hence knowing the presence of other habitats within 2–5 km helps understand the likely impacts of developments. For species such as bats, the minimum search area for records is 2 km, but it is up to 10 km for large sites or where there are SSSIs or Special Areas of Conservation (SAC) relevant to bats (Collins 2016). For wide-ranging species such as birds of prey (Hardey et al. 2013) or Otter (Macdonald and Tattersall 2001), similar issues exist. In England, the NE MAGIC database (https://magic.DEFRA.gov.uk/) will provide much of the mapped material expected to accompany a planning application. None of this is considered in the 2022 SA. Appropriate distances are part of the considerations to include in the desk element of a PEA (see above).

5.1.3 Site status

Just as species can have protected status, so may sites. In addition to the standard SSSI, SAC and local nature reserves (LNR) there are other designations attached to land that will also have resonance in planning terms. These will need to be considered as part of the potential impacts of a proposed development. These should be picked up in the PEA and the desk search element.

5.1.4 Species status

The PEA should have picked up the range of species in and around the site and will have flagged those protected species covered by the SA. These species require special care and coverage. Detailed and caveated survey data and historical data will be expected as part of the examination of potential status and impacts in any planning submission. In addition to protected species, there may well be species of local importance or restricted ranges too.

5.1.5 Data age and sources

All data are not equal. As species numbers and ranges fluctuate, so the meaning, and reliability, of data will change through time. Collins (2016) recommends that bat data should be no older than the previous field season. Both CIEEM (2019) and BS 42020 caution against uncritical use of data. In practical terms, as data age, so they become unreliable (CIEEM 2019a,b) and may need updating after between 18 and 36 months. After three years, reports based on 'old' data are unlikely to be valid and will need completely updating (CIEEM 2019b). Note that for NatureScot, protected species data will need to be updated within two years if an application drags forward (e.g. Badgers https://www.nature.scot/doc/standing-advice-planning-consultations-badgers). The need for updating is greater for those locations where there are known habitat changes and/or where highly mobile species are involved (CIEEM 2019b).

Where desk data are being used, there is a clear need to be aware of the limitations of such data, especially as many datasets are fragmentary, and there is often little context provided to datasets summarised in LRCs or in specialist groups' data trawl replies. The same applies to NBN data upon which NE is keen to rely. NE's TIN069 critique of LRC datasets for use in windfarm assessments is especially damning (NE 2010).

Field datasets commissioned to accompany an application should follow BS 42020 closely, and properly describe what was done and not done so that the merits and limitations of the dataset are clear. That is not always the case (Reed 2021).

5.1.6 Field data collection methods and reliability

If desk data are often less than robust, it is important that data collected for protected species are fit for purpose. It was noted earlier that the old NE SA that was replaced in January 2022 was ill-suited as guidance for either LPAs or applicants. The 2022 SA is far poorer still. This means that field data collection will need to rely on distillations of guidance, such as those in the tables below and the detailed methodological texts for individual species or species groups. These methodological texts require levels of experience, competence or capacity missing from most LPAs. Because of this, it is important that pros and cons for each group are summarised for LPA benefit. Some of the key issues are noted in the following sections. Note that both the PPP and the PSD indicate that LPAs can refuse applications if the survey data are found wanting – requiring a new set of surveys to be undertaken.

5.1.6.1 Objectives stated

Surveys can range from simple walkover surveys to get a feel of the site and surroundings – as often used in precursors to PEAs or in more detail in PEAs themselves. These 'extended phase 1 surveys' will produce very different results from those intended to find and identify individual species groups. There is a temptation in some PEAs to assume too much from a first visit (e.g., make an assertion about a summer bird community at Smithy Wood on the basis of a cursory visit in mid-winter; (Woodland Trust 2016)).[3] If there are concerns about the possibility of protected species occurring, that is the role of a detailed follow up in a phase 2 survey. That applies across a range of taxa. The phase 2 survey is usually undertaken to inform an EcIA. All protected species surveys will need the detailed methods set out in Tables 5.1–5.16.

Some standard survey techniques, such as the bird method Breeding Bird Survey,[4] are designed to help produce indices, but they are not to be used for species estimates or habitat use assessment, emphasising that surveys are objective-led.

5.1.6.2 Timing a visit

Single visits to a site are indicative and cannot be expected to be categorical. The phase 1 guide (JNCC 2010) is clear on the problems of inference from visit timing. Botanical assessments in late winter or early spring will be very different from those later in spring or summer (JNCC 2010), and it is quite possible to chart changes over the course of a summer as plant species appear and then disappear, meaning that single visits in each period would miss many of them.

Timing issues apply equally to animals. Visits in winter will reflect very different avifaunae from those in spring or summer. Survey methods for individual bird species, such as nightingales, will emphasise different times of the season for visits and different times within the day according to season. Assessments of a pond in mid-winter will produce different amphibian results from one in late March.

---

[3] https://www.woodlandtrust.org.uk/protecting-trees-and-woods/campaign-with-us/smithy-wood/

[4] https://www.bto.org/our-science/publications/birdtrends/2019/methods/breeding-bird-survey

5.1.6.3 Time of visit

Just as the time of year significantly impacts detectability, so the time within the day does too. Individual guidance emphasises the effects of time of day on species as varied as bats (Collins 2016), birds (see above; Gilbert et al. 1998), mammals (Hill et al. 2005), reptiles and amphibians (Gent and Gibson 2003) and insects (Drake et al. 2007). Getting this right, and documenting this, is critical. This is one of the limitations frequently ignored in planning submissions.

5.1.6.4 Number of times to visit

Just as time of year and time of day affect detectability and behaviours (Hill et al. 2005), so the number of visits needed to produce meaningful data will also vary too. This will be determined by objectives (see above) and hoped-for reliability. Single visits will produce radically different data from multiple visits and will influence how the data are used and interpreted. For many species, there will be a minimum number of visits required to understand which areas are used, and how. This applies for example to birds, amphibians and reptiles, and Water Voles (Gilbert et al. 1998; Gent and Gibson 2003; Strachan et al. 2011; Hill et al. 2005).

5.1.6.5 How long to survey?

Depending on the species and the objective, some parts of survey works may require a month or so to provide requisite data. For example, undertaking Badger bait marking surveys usually requires at least 21 days of regular visits to detect and determine territory use (NS Badger best survey practice 2020). Being aware of the actual time needed for an aspect of species surveys is a key to assessing whether or not surveys are adequate.

5.1.6.6 Weather

No matter if the visits are at the correct times of year or day, if the weather is unsuitable there will be a serious impact on the quality and quantity of data collected and how they may be interpreted. Bats do not fly in cold, wet windy conditions (Collins 2016), and this equally affects butterflies and song detectability in birds amongst others (Hill et al. 2005). Reptiles are strongly affected by sun and temperature (Gent and Gibson 2003). Yet weather is frequently ignored as a limitation in planning submissions, even when examination of records shows them to be a major problem (Reed 2021) . Avoiding poor conditions is critical to provision of good-quality data needed for planning evaluation.

5.1.6.7 Light conditions

The majority of certain species groups, such as moths, will only be found with any frequency in poor light or dark conditions. Others, such as bats, typically only appear as the light goes, and the time around dusk or twilight may be critical for certain species of bat (Collins 2016) and birds such as nightjars (Gilbert et al. 1998). Unless night vision technology is used, visual acuity declines for detecting most species as the light becomes poorer. As it does, efficiency is greatly reduced. In addition, practical safety issues mount as light fades (Hill et al. 2005). Looking for fish or crayfish demands good light conditions (see tables below).

5.1.6.8 Distance

Distance affects detectability: animals become visually hard to detect with distance, and noises attenuate quickly. This means that birds and insects may be missed, mammals overlooked, and sounds lost with distance. For groups such as bats, where detectability distances on most detectors are very limited (Adams et al. 2012), many remotely recorded datasets are compromised by being too far away for effective detection. For species such as Brown Long-eared Bat *Plecotus auritus*, the very short distance over which calls can be detected means that field-based detection is equally problematic for confirming presence as absence (Russ 2012).

For groups such as birds, windfarm guidance places upper distance limits on detectability, and many transect methods for various groups impose an upper distance limit between transects to minimise loss of individuals as they fall out of vision or earshot.

5.1.6.9 How far to survey

Every site has a wider area that provides its context. It is clearly unusual to cover just a pinpoint and ignore the rest of the surroundings. To reflect this, many methods for species look at the site and suggest a distance beyond the site boundaries that would also need covering on survey visits. The distances vary between species, as does the sort of information collected. This may be as far as 250 m for Otter (SGN), 500 m for Great Crested Newt (Sewell et al. 2013), Water Vole on large developments (Dean 2021; NS 2020a), 500 m for windfarms (SNH 2017) or 1 km for Badger (NS 2020b).

5.1.6.10 How many years of survey data

For many applications, there is often only one year's worth of data provided. Given some of the issues noted above, a single year's sample may or may not be misleading. For that reason, windfarm studies require at least two years of data (NE 2010; SNH 2017).

5.1.6.11 Limitations

With so many factors that need assessing, it is unsurprising that not all will have been met on any one occasion. This means that limitations might be expected to be one of the biggest and most critical sections of a planning submission (Reed 2021) and should appear in all protected species data records and interpretation (BSI 2013).

5.1.7 Licences and competencies

Most, but not all, protected species require licences to survey, disturb or monitor them. All guidance provides basic facts on licences and requirements. When applying for a licence, an applicant is expected to have met a series of competency tests or thresholds. It does not follow that meeting all competencies will be equivalent to licence holding. Any well-trained person may be competent but that differs from being a licence holder. For a range of mitigation and disturbance activities, it is critical that the ecologist has the requisite licence(s). Failure to do this would render the development illegal and put the ecologist at risk professionally (by failing to meet CIEEM standards) and legally. Documentation of licenses is a basic part of a submission where many activities will be otherwise illegal.

5.1.8 Understanding impacts

In order that the mitigation hierarchy can be applied, it is critical that the impacts are understood, both locally and as part of a cumulative assessment. This requires robust quantitative data and errors and limitations to be explained, and substantiated, in plain view.

5.1.9 Avoidance, mitigation and compensation

These can either be aspirations or facts (Hunter et al. 2021). Where mitigation programmes are to be used, it is important that areas of habitat are suitable: in kind and potential supporting capacity. All, including compensation, must avoid hoping for suitable habitats. To be effective, they must be real and suitable habitats, so that when effects are felt there is no developmental lag before habitats can be used. This needs to be part of any SA guidance.

5.1.10 Monitoring

In order to assess if there is no impact, or net positive gain, linked to an application, or even if the conditions attached to the development have been met, it is important that there is monitoring guidance linked to any species material being used by an LPA. This should cover the duration, methods – usually the same as the survey methods in the application – and confirmation of how this will be financed and acted upon if feedback shows that problems are developing. This should all sit within a validated site management plan.

5.2 Progressing

If these basic requirements are understood, and adhered to, then there is a clear opportunity to progress and evaluate the potential merits and risks of the proposed development. But if data collection is of uncertain quality, and suggestions in the mitigation and compensation measures are weak at best, then the LPA is placed in an invidious position, and with it the hope of NNL or at least 10% BNG being attainable. Note that the BNG concept is habitat based (Zu Ermgassen et al 2021), so that it assumes that protected species will come along as part of a habitat package; that is simplistic. Protected species are still subject to effective species-focused initiatives and legislation (Panks 2022).

The following tables for species and groups indicate some of the issues that LPAs should look out for in their evaluation of protected species data as they appraise applications.

5.3 Protected species guidance for planners in tables

Table 5.1 The basic considerations to include for any protected species or habitat data to be used in a planning applications presented to an LPA

Species	Notes
Site name and status and zone of influence	Site status will be important for level of protection, including protected species that will affect an application. Status and location of designations within 5 km to be considered; 10 km if bat-based designation.
Identified in pre-application discussions	Pre-application discussions will normally include all aspects of biodiversity – especially Protected Species.
Picked up in Preliminary Ecological Appraisal	Preliminary Ecological Appraisal will include desk surveys from Local Record Centres and specialist groups as well as field surveys. Distances extend beyond site boundaries. Once a PEA picks up the species, a more detailed EcIA will normally be needed.
Protected species in and around the site	All need to be identified for immediate or secondary or cumulative impacts.
Age of data attached to the application	Desk-search data age quickly, and data reliability reduces accordingly. Desk data are patchy and only indicative, not absolute. Field-collected data are out of date within two years, and all by three years.
Field collection methods	These should clearly reference the standard(s) used and describe how these were applied or problems that limited their application. Unsupported use of professional judgement is unacceptable and invalid.
Objectives of surveys	Methods differ in their suitability according to purposes. These need to be clearly stated and have a description of validity.
Survey timing	Timing within the year affects the use of phase 1 surveys and extended phase 1 surveys and is a limitation to consider seriously in evaluating results. It also affects individual species' survey methods.
Timing of visits in the day	Timing within the day is a critical issue for detectability for many species. A poorly timed visit will severely bias results.
Visit frequency	Visit frequency for survey purposes is specified, and individual techniques will have their own needs. Failure to meet these is a limitation to be specified and will affect reliability and basis for assessment and monitoring baselines.
Survey duration	Individual visits will need a minimum time and some techniques (e.g., bait marking) will require specific time periods.

SURVEYS FOR PROTECTED SPECIES: WHAT THE LPA MIGHT HAVE ORDERED | 95

Species	Notes
Weather	Weather conditions (wind, rain, temperature) will all affect detectability/emergence significantly and are clear limitations and cannot be covered by unsubstantiated assertions of professional judgement. Poor conditions may also prejudice the wellbeing and survival of many species if disturbed.
Light	Light is a key factor affecting behaviours, detectability and survey efficiency. This applies equally to aquatic surveys as to terrestrial surveys.
Distance	Distance from detectors or observers is a critical factor in determining detectability and survey efficiency and needs validating. Transect methods specify maximum distances between lines/observers. Distance is also a factor in survey requirements for detecting contiguous populations.
How far beyond development boundaries	Species occur beyond site boundaries, and methods require different distances according to species.
Years of data	For some species and development types, several years of data are required. This also applies where limitations in a sample year preclude meaningful interpretation.
Limitations	All survey attributes are rarely met. A clearly based assessment of limitations is central to all impact assessments. Professional judgement needs substantiating to be used; without it, it is just personal opinion.
Licences and competencies	For almost all protected species, disturbance will require a licence to proceed, and the ecologist will need to be highly experienced in that field. Both need stating in an application. CIEEM indicates expected competencies for almost all protected species.
Impacts: single and cumulative	Understanding impacts depends on data, data quality and context. Immediate and cumulative impacts will need to be considered, especially where species are considered in district scales together.
Avoidance	Avoidance presumes reliable datasets, so that the implications of avoiding a site are clear.
Mitigation	Mitigation requires detailed impact assessments, and these require data that are fit for purpose and are robust.
Compensation	As above.
Monitoring	Monitoring is part of a detailed plan and requires comparable methods and reliability of data for a specified period, with finance, plus review periods and responses. Monitoring without context is just a set of observations, with no linked actions. That has little basis for biodiversity net gain evaluation.

Table 5.2 Species guidance: Badger

Species: Badger	
Site name and status	Badgers occur on designated and non-designated land. In England, site status can be gleaned from Natural England's MAGIC website and is a basic part of a PEA desk search (CIEEM 2019a). It should also be listed in the application.
Zone of influence	Badgers occur on all types of land and a wide range of habitats (75% of setts were in woods, hedges and scrub (Delahay et al. 2008)) and forage widely, commonly up to 1–2 km from the sett (Harris et al. 1989). This means that any application needs to include more than the immediate boundaries (NS 2020b), and time will be required to provide detailed coverage of areas before submission. Locations of recorded setts may also come from desk survey data, but many LRCs limit data release for Badgers.
Identified in pre-application discussions	Pre-application discussions will normally include all aspects of biodiversity, especially protected species which includes Badgers. They are covered under the 1992 Badger Act. Most LPA discussions will refer to validation checklists, and these will include Badgers. If they are not in the discussion, this should be queried by LPAs.
The species should be picked up in Preliminary Ecological Appraisal	Preliminary Ecological Appraisal will include desk survey data sought from LRCs and specialist groups – in this case county mammal groups – as well as field surveys. Distances extend up to 1 km beyond site boundaries. Extended phase 1 surveys should pick up setts – especially if in the February to April or October periods. Coverage for surveys would be about 2 km^2 per day (NS 2020b).
Protected species in and around the site	As Badgers may well range into and beyond the application site, the preliminary walkover field search is a key first stage of potential impact assessment and should use methods set out below to confirm range and site(s) used. Coverage must extend 1 km beyond site boundaries, even for a large site, as this may well form part of one or more external territories.
Age of data attached to the application	Field data are out of date within 2 years. NatureScot (2020a) requires resurveys if data are older than 24 months. As a subset of age, Badger setts may be left unused for several weeks, so once found their status will need to be checked again to assess usage (NE 2009). Single visit Preliminary Ecological Appraisal surveys need caveats on status.

SURVEYS FOR PROTECTED SPECIES: WHAT THE LPA MIGHT HAVE ORDERED | 97

Species: Badger	
Objectives of surveys	The objective of the walkover survey in the Preliminary Ecological Appraisal (PEA) is to provide outline evidence of presence; absence takes more time. If Badgers are suspected, then this should be picked up in a phase 2 requirement for more detailed coverage. Bait surveys and other surveys will be needed to assess home ranges and the basis for potential disturbance and licensing requirements. If the focus of the surveys is to establish main and satellite set use, then at least a month of detailed survey work is needed for bait marking studies. Any licensed exclusion work will need additional time allowances to confirm status and alternative use areas. Survey methods should be explicitly stated so that they can form the basis for any post-development monitoring. The walkover survey will also produce a map of phase 1 habitats (JNCC 2010), which will provide an indication of the main habitat types in and around the site and will be needed to assess habitat preferences and the potential effects of habitat loss by the proposal, and the secondary effects of fragmentation and isolation of habitats used.
Field collection methods	In order to ensure that the site and areas around are fully assessed for the presence of Badgers, both sides of all hedges, boundaries and all woodland within 1 km of the site should be surveyed, at the appropriate time, using the methods set out below. Special focus should be on hedges on slopes where setts may be on the downside (92% of setts were found in slopes (Mammal Society 1963)) but are not necessarily visible on the upside. A range of signs can be used to suggest presence: 1. Faeces: Badgers use small pits to deposit faeces; some are near (<50 m) setts, and concentrations of these latrine areas are often on territory boundaries. Detectability reduces over the summer with vegetation growth. Faecal composition will alter with seasons, helping show recency. 2. Scratching posts: bases of trunks, fallen branches or posts are used for territory signs or paw cleaning. Used posts do not show signs of moss or cobwebs. 3. Hair traces: hair is often caught on thorns, barbed wire or rough projections as Badgers pass. 4. Footpaths: networks of paths lead out from setts and will produce well-worn tracks, often flattening vegetation below surrounding ranker growth. 5. Footprints: these will be found in damp areas or in soft bare earth after rain- including outside of setts. Very much weather dependent. 6. Snuffle holes: scrapes where Badgers search for food items. Less obvious in rank vegetation. 7. Day nests: bundles of usually dry/drying grass where animals sleep above ground. 8. Setts: single or multiple holes interconnected underground and joined by paths to feeding/latrine areas and rest of territory.

(continued)

Table 5.2 Species guidance: Badger (*continued*)

Species: Badger	
	Any report should confirm sett activity status by: A) well-used: some of 1–6 plus, well-worn entrances with freshly excavated soil or bedding. B) partially used hole/holes: leaves/twigs in entrance; moss/other plants growing in/around entrance. C) disused: holes blocked with earth/vegetation and would need effort to use and open. The potential effect on setts from a proposed development also depends on the type, which would also need to be known and documented with reasons: A) main: several holes with large active spoil heaps and paths to and from holes. B) annexe: usually <150 m from main sett; its several holes may be used occasionally. C) subsidiary: >50 m from other setts, with no obvious linking paths and intermittent use. D) outlier: little spoil, no linking paths, often evidence of fox/rabbit use. Establishing the status and use of the sett is critical for impact assessment, along with the territory area used by Badgers from those setts. The most widely used method of assessing the presence of Badger groups, and the range/territory used, is coloured bait marking. It is used to establish limits of social group territories – essential when setts are within 1 km of a proposed development area. Bait marking works by placing indigestible coloured plastic markers in food (usually a mix of peanuts and syrup) by a main Badger sett which is then eaten and digested. The main latrines away from the sett, normally on the territory boundary, are used as part of territorial marking, so being able to see which social group is using which is important. By inspecting latrines for coloured markers, it will be possible to see which areas and the range of locations that a group uses. For maximum effect, the study needs doing in spring and/or autumn.
Survey timing	Badgers are active all year round. The optimal time for Badger survey is the peak territorial period of February to April when vegetation is also low, with a second period in early September to the end of October. As vegetation increases, so detectability of the main signs become less obvious, even if Badgers are present. If bait marking is used, spring is best, with a second autumnal period in September to October (NS 2020b). Using 'wrong' times is a limitation to data reliability. Any exclusion work (see licensable below) must take place within the period July to November inclusive.
Timing of visits in the day	If the objective is to see Badgers, emergence times vary across the year and with weather and temperature and disturbance (Neal and Cheeseman 1996). If detection is to be used to confirm status, then camera traps are best used over a period of several weeks, to allow for limitations.

Species: Badger	
Visit frequency	Bait marking is the main way of assessing Badger range and status. A valid study requires daily checks and bait supply for at least 21 days, although this may be longer depending on weather (NS 2020b), with time then needed for data analysis. If 2–3 setts and territories are involved, this is a major time investment. Failure to properly undertake the survey is a limitation to be specified and will affect reliability and basis for assessment and monitoring baselines.
Survey duration	At least 21 days for bait surveys in the correct months, and most likely 28 days (NS 2020b).
Weather	Weather conditions (wind, rain, temperature) will all affect detectability/emergence significantly and are clear limitations and cannot be covered by unsubstantiated assertions of professional judgement.
Light	Light is a key factor affecting behaviours, detectability, and survey efficiency. For Badgers, use of trail cameras will help assess both levels of use, emergence times and also social group composition. If used, cameras should cover all holes in the main sett and those thought to possibly be used as annexes.
Distance	Searches 1 km from the site boundary will be needed to indicate the potential or actual use of a territory.
How far beyond development boundaries	For Badgers, where several social groups may be involved, 1 km beyond the site limit is needed. This also reflects the potential for the site to be used by Badgers that have setts in suitable areas offsite and which would be possibly affected by changes in the development area, as well as those on the site itself.
Years of data	At least one year of data is needed. More may be needed where data limitations affect reliability. This also applies where limitations in a sample year preclude meaningful interpretation.
Limitations	All survey or weather attributes are rarely met. A clearly based assessment of limitations is required. Because of the possibility that the public may mistake bait stations as poison bait (NS 2020b) police and local wildlife authorities (including potentially Natural England) should be advised, so that bait marking is not cut short, limiting its value. Any assessment of limitations needs data to support conclusions; professional judgement is a personal opinion without proper data.
Suitable licensed and experienced surveyors	Surveyors are expected to be competent (CIEEM 2013f) have practical experience, with at least two accompanied surveys (by a surveyor with experience of at least 20 sites) of sites where the majority of Badger field-signs are present. Where the setts have to be disturbed, surveyors need to hold a Badger licence.

(continued)

Table 5.2 Species guidance: Badger (*continued*)

Species: Badger	
Determining impacts	Assessing impacts will require knowing the locations of setts, the pattern and use of territories and preferred habitats. A Badger protection plan will need to include information on loss/disturbance to main or annexe setts, loss of foraging habitats and implications of habitat fragmentation as the basis for mitigation. All of this should have been determined by field surveys above.
Impacts: single and cumulative	For social groups such as Badgers, loss of habitat and foraging areas or fragmentation may act individually or cumulatively on one or more territories. Disturbance may have similar effects over time.
Mitigating properly assessed potential impacts	Basic principles: development should avoid most heavily used habitats; there should be identification of suitable areas for mitigation by developing similar habitats but with access and low disturbance likely; linear features should be retained; if necessary, tunnels or similar should be installed to allow access between areas; traffic calming should be included in Badger areas; and lighting in or near areas used by Badgers should be minimised. In building times, machinery should be off two hours before sunset, light should be directed away from Badger use areas and trenches sealed daily, water sources should be sealed and pipes capped daily.
Licensable activities	Disturbing a sett requires a licence. Any annexe setts should be monitored for current status: Badger use will vary across an often-prolonged development. If setts need to be disturbed, a licence will be needed. Sett closure would require a minimum of 21 days exclusion after emptying via a one-way door. Only when it is certain that the sett is empty can destruction begin. In extremis, an artificial sett may need construction.
Monitoring	If mitigation is planned, then it is normally required to be monitored. Objectives will need to be stated (e.g., determining use of fencing or underpasses or state of Badgers excluded from destroyed sets: this may include use of expanded setts or artificial setts – methods will mirror those in pre-development period). Timing will meet normal survey methods and frequency, and require several periods of observation per year. A minimum of a well-budgeted two years of post-development monitoring will be needed.
Summary	A Preliminary Ecological Appraisal is required, with a sett search 1 km beyond boundaries. Detection is influenced by time of year: best February to April and September to October. A range of signs are used to determine status. Main and other setts need to be differentiated and checked. For territorial assessment bait-marking, a minimum of 21–28 days is required. Mitigation needs to assess habitat use, limit isolation and fragmentation and requires two plus years of post-licence monitoring to assess effectiveness.

Table 5.3 Species guidance: Natterjack Toad

Species: Natterjack Toad	
Site name and status	Natterjack Toads occur on designated and non-designated land. In England, site status can be gleaned from Natural England's MAGIC website and is a basic part of a PEA desk search. The walkover element of a PEA will indicate suitable habitat and possible evidence of occurrence. As a European Protected Species and a species covered under section 41 of the NERC Act (2006) it should also be listed in the application.
Zone of influence	Natterjack Toads are now found almost entirely in three habitat types: sand dunes, saltmarsh and lowland heath. The key habitat requirements are shallow (often ephemeral) warm ponds for breeding and open, sandy terrestrial habitats for foraging, dispersal and hibernation. Natterjacks are relatively highly dispersing (Sinsch et al. 2012), with annual movements up to 4.4 km and core ranges of within 600 m of a breeding site. On that basis, the zone of influence needs to be ≤1 km.
Identified in pre-application discussions	Pre-application discussions will normally include all aspects of biodiversity – especially protected species, which includes Natterjack Toad. Most LPA discussions will refer to validation checklists, and these will include toads in districts/counties that hold them. If they are not in the discussion, this should be queried by LPAs.
The species should be picked up in the PEA	The PEA will include desk survey data sought from Local Record Centres and specialist groups – in this case county herpetology groups – as well as field surveys. Distances extend up to 1 km beyond site boundaries. As NE (2022) notes, absence of Natterjacks from a desk search is not the same as site absence. Extended phase 1 surveys should pick up potential natterjack habitat within known ranges.
Protected species in and around the site	After using the PEA desk search as a screening for records in and around the site, the preliminary walkover habitat mapping and field search is a key first stage of potential impact assessment and should use methods set out below to confirm status (NE 2015, 2022). Coverage should extend ≤1 km beyond site boundaries to evaluate the potential for recolonisation if the breeding site within the potential development areas is potentially impacted or habitats are lost. Maximal distance between ponds is ≤2.25 km (Sinsch et al. 2012).
Age of data attached to the application	Field data are out of date within two years (CIEEM 2019b; NE 2022a).

(continued)

Table 5.3 Species guidance: Natterjack Toad (*continued*)

Species: Natterjack Toad	
Objectives of surveys	The objective of the walkover survey in the PEA is to provide outline evidence of presence or potentially suitable habitat on or near (NE 2022a) the site. Note that toads will use a core zone of at least 600 m beyond the breeding ponds, with at least 50% using a radius of around 1150 m (Sinsch et al. 2012). The main objective of a correctly timed detailed phase 2 survey is to establish the presence of Natterjacks, indication of areas used and possibly abundance. Several methods are available and more than one is needed to establish site populations and structure (NE 2022a). Survey methods should be explicitly stated so that they can form the basis for any post-development monitoring. The walkover survey will also produce a map of phase 1 habitats (JNCC 2010), which will provide an indication of the main habitat types in and around the site and will be needed to help assess habitat preferences and the potential effects of habitat loss by the proposal and the secondary effects of fragmentation and isolation of habitats used. A phase 1 walkover in autumn/winter, while poor for detailed vegetation (JNCC 2010) assessment, would be suitable to assess basic habitat suitability.

Field collection methods	In order to ensure that the site, and areas around it, are fully assessed for the presence of Natterjacks, all suitable areas within 1 km of the site should be surveyed, at the appropriate time, using methods set out below (Beebee and Denton 1997). A range of methods can be used, including (a) qualitative data and (b) quantitative data.
	A) Qualitative:
	  1. Calling males: early April to early May; the period from dusk to midnight on warm, damp, nights, males call vigorously. As not all males do this, and not all males are at breeding sites, this is indicative only.
	  2. Refugia use: between April and October, toads will make use of natural refugia. Artificial refugia (such as carpet or tin) can increase the chances of finding toads, but is not guaranteed.
	  3. Torch searches. Toads hunt at night, and on warm, damp nights between April and October in sites with few refugia, they may be spotted by strong torch light. This is badly affected by dry periods.
	B) Quantitative:
	  1. Spawn string count: Natterjacks spawn separately from each other. Visits to prospective breeding ponds should be at least once a week from at least early April to early June and involve counting spawn strings. Spawning may be delayed in dry spells, and sped up after rain. With an error term of around 5%, string counts indicate the number of adult females and if doubled suggest the total adult population. This is less categoric in very dry springs where up to 100% of females fail to breed.

SURVEYS FOR PROTECTED SPECIES: WHAT THE LPA MIGHT HAVE ORDERED | 103

Species: Natterjack Toad	
	2. Toadlet counts. Counting emerging toadlets is a direct measure of breeding success and indication of future adult population size. Metamorphosis is between mid-May and July, peaking in June. Ponds should be checked weekly for tiny toadlets. Once seen, ponds should be visited every day for 14 days. A week's gap at that stage can completely miss evidence of successful metamorphosis. Weather (warm damp is ideal) affects dispersal rates of toadlets.
Survey timing	Survey timing in early April to early June is critical for spawning surveys. For toadlet emergence, mid-May to July is the key period. For adult checks, refugia can be seen in spring to autumn, and the same period should be used for night-time torch checks.
Timing of visits in the day	Torch checks are best done from dusk to midnight. Calling male checks are done from dusk to midnight. Refugia use is in the daytime, and toadlet checks are best in the daytime for counting and to avoid trampling at night.
Visit frequency	Spawning requires at least one visit per week and toadlet counts at least once daily for 14 days after first toadlets emerge.
Survey duration	For adult call counts, timing is early April to early May; for toadlets, there should be a 2–3 weeks minimum survey.
Weather	Weather conditions (wind, rain, temperature) will all affect activity of adults, spawning rate and emergence of toadlets. Prolonged dry periods may stop spawning.
Light	Light is a key factor affecting calling and hunting activity and toadlet dispersal.
Distance	Searches up to 1 km from the site boundary will be needed to indicate the presence of toads and suitable alternative breeding areas. Dispersal may be up to 2–4 km (Sinsch et al. 2012).
How far beyond development boundaries	As toads occur at low density, and do move up to >1 km on dispersal, 1 km beyond the site limit is needed.
Years of data	If the objective is to establish presence, then a single year may suffice. If it is to confirm absence, then a minimum of two will be needed. To establish population status a longer period of 5 or more years will be needed.
Limitations	Natterjack behaviour is strongly affected by weather (temperature, rain, humidity and runs of dry days) and daylight. Choosing correct weather, times of day and frequency of visits according to parameters being checked will impact data value and impact assessment claims.

(continued)

Table 5.3 Species guidance: Natterjack Toad (*continued*)

Species: Natterjack Toad	
Suitable licensed and experienced surveyors	Surveyors are expected to have practical experience (CIEEM 2013g), under the supervision of a licensee and to have: A) performed surveys of Natterjack Toads across all life stages (spawn to adult), including capture and biometric measurement, in compliance with the Natterjack Toad Conservation Handbook (Beebee and Denton 1997); B) surveyed at least 10 different breeding ponds, ideally at several different sites, under supervision of a Natterjack Toad licence-holder and survey over at least five different occasions, spanning the breeding season from April to July. A survey licence is required for any survey work using techniques likely to disturb Natterjack Toads at any site where there is a reasonable likelihood that they are present (e.g., terrestrial searches or netting).
Determining impacts	Assessing impacts will require knowing the locations of spawning ponds, potential adult population sizes and areas used and preferred habitats. Negative impacts will include habitat loss, breaking of linear connections between areas, fragmentation and increased levels of post-development habitat disturbance. The simplest option is to avoid disturbing known populations. This includes indirect effects on water tables, which are critical for spawning, tadpole growth and toadlet survival.
Mitigating properly assessed potential impacts	Basic principles should include: developments should avoid breeding sites or altering water levels; identifying heavily used habitats and areas; identifying suitable areas for mitigation by developing similar habitats but with access and low disturbance likely and well in advance of any planned development so that habitats can become suitable for toad use. Hibernation shelters should be considered, along with managing water areas under condition of low nutrient addition or shading. Any habitat compensation will need to be well established before use by toads. Translocation must be a last option and has associated risks of disease occurrence.
Licensable activities	Disturbing Natterjacks, or areas known to be used by them, requires a licence. Any habitat alteration will require a Natterjack Toad mitigation licence issued by Natural England. Any translocation would also require a licence.
Monitoring	If mitigation is planned, then monitoring is normally required. Objectives will need to be stated. Timing will meet normal survey methods and frequency and require several periods of observation per year. A minimum of a well-budgeted two years of post-development monitoring will be required.

Species: Natterjack Toad	
SUMMARY	A PEA is required, with a searches up to 1 km beyond boundaries. Detection is influenced by time of year: best is April to October. A range of methods is used to determine status. For status assessment, breeding surveys and toadlet checks are required. Mitigation needs to assess habitat use, limit isolation and fragmentation and requires 2 or more years post-licence monitoring to assess effectiveness. Missing key times in the survey period may render data useless. Weather is a key factor influencing survey efficacy.

PEA = Preliminary Ecological Appraisal; NE= Natural England

Table 5.4 Species guidance: Hazel Dormouse

Species: Hazel Dormouse	
Site name and status	Dormice occur on designated and non-designated land. In England, site status can be gleaned from Natural England's MAGIC website and is a basic part of a PEA desk search. The walkover element of a PEA will indicate suitable habitat and possible evidence of occurrence. As a European Protected Species and a species covered under section 41 of the NERC Act (2006), it should also be listed in the application.
Zone of influence	Hazel Dormice occur across a range of habitats, including broad-leafed woodland, coppice, coniferous plantations and broadleaf scrubland (Mathews et al. 2018) and will slowly disperse across the landscape through hedges and scrubby edge communities. Woodlands used are typically >20 ha with a wide range of growth stages, with significant amounts of understorey, particularly those ≤500 m from the nearest suitable habitat (Bright et al. 2006). This means that any application needs to include more than the immediate boundaries and 1km around (Bright et al. 2006). The wide range of habitats used means there is no simple 'typical' habitat (Bright et al. 2006; Juskaitis and Buchner 2013).
Identified in pre-application discussions	Pre-application discussions will normally include all aspects of biodiversity, especially protected species which includes Hazel Dormice. Most LPA discussions will refer to validation checks, and these will include dormice in districts/counties that hold dormice. If they are not in the discussion, this should be queried by LPAs.

(continued)

Table 5.4 Species guidance: Hazel Dormouse (*continued*)

Species: Hazel Dormouse	
The species should be picked up in PEA	PEA will include desk survey data sought from Local Record Centres and specialist groups – in this case county mammal groups – as well as field surveys. Distances extend up to 1km beyond site boundaries. As NE (2022a) notes, absence of dormice from a desk search is not the same as site absence. Extended phase 1 surveys should pick up potential dormice habitat (but this is varied – see above), and where hazel occurs evidence of distinctively gnawed hazel shells is enough to confirm Hazel Dormice (Bright et al. 2006). Nests are hard to find. Finding either feeding signs or nests, or, if suspected after a walkover and tube sampling, this status will be enough to protect the site in principle. Nests are best found in autumn/winter in low shrubs and are round grapefruit-sized balls of full-sized tree leaves and bark, with no obvious ways in or out.
Protected species in and around the site	After using the PEA desk search as a screening for records in and around the site, the preliminary walkover habitat mapping and field search is a key first stage of potential impact assessment and should use the methods set out below to confirm status (NE 2015c). Coverage should extend <1 km beyond site boundaries to evaluate the potential for recolonisation if the site is potentially impacted or habitats lost.
Age of data attached to the application	Field data are out of date within two years (CIEEM 2019b). NE (2022c) will accept dormice data up to three years old if habitats have not changed significantly, but there is no guidance on what 'significantly' means, and it requires definitive data, so two years is best.
Objectives of surveys	The objective of the walkover survey in the PEA is to provide outline evidence of presence; absence takes more time. If dormice are suspected, then this should be confirmed in a phase 2 requirement for more detailed coverage by a licensed dormouse surveyor. The main objective of a detailed phase 2 survey is to establish the presence of dormice and an indication of areas used and possibly abundance. This will also include areas beyond the site (NE 2015c, 2022c). Several methods are available. Survey methods should be explicitly stated so that they can form the basis for any post-development monitoring. The walkover survey will also produce a map of phase 1 habitats (JNCC 2010), which will provide an indication of the main habitat types in and around the site, and will be needed to assess habitat preferences and the potential effects of habitat loss by the proposal and the secondary effects of fragmentation and isolation of habitats used. A phase 1 walkover in autumn/winter, whilst poor for vegetation (JNCC 2010), might be suitable to detect some signs of dormice. See below for details.

Species: Hazel Dormouse	
Field collection methods	In order to ensure that the site, and areas around, are fully assessed for the presence of Hazel Dormice, hedges, boundaries and all woodland within 500 m to 1 km of the site should be surveyed, at the appropriate time, using the methods set out below. A range of signs can be used to suggest presence in order of preference: 1. Nuts. The examination of hedgerow bottoms for the distinctively eaten hazel nuts, where hazels occur, can show presence. Clearly, this is best in mid- to late autumn when nuts are most available and not taken away or degraded. By late winter onwards, detectability is reduced, and difficult to almost impossible once spring herb growth starts (Bright et al. 2006). At least five 10 × 10 m quadrats in areas with hazels are needed, more so where squirrels are present and compete for nuts. Note that failure to find gnawed nuts is not the equivalent of absence. 2. Boxes and nest tubes. The objective of box and tube surveys is to detect dormice. The functions of the two differ. Boxes are used primarily for nesting and breeding, while nest tubes are mostly used for short-term shelter and day nesting (Bright et al. 2006; Chanin and Gubert 2011). Of the two, boxes are the only approved way of monitoring long-term dormouse populations. In the short-term, boxes may take a while to be used, while uptake of tubes is faster (Bright et al. 2006; Chanin and Gubert 2011) as dormice may use up to three different day-nesting sites in a week. Nest tubes are an effective way of sampling in the absence of hazel in March to November. 2a. Nest tubes. A minimum of 50 tubes, spaced at 20 m intervals, should be installed by the end of March on the underside of branches up to 2 m above the ground, allowing visual access. As dormice use different heights of woodland and scrub across the season, the index detectability probability score of finding signs of nests or dormice themselves varies from month to month. The highest are May, August and September. A minimum detectability score of 20 is needed (Bright et al. 2006). Monthly or bimonthly inspection is required. 2b. Nest boxes. Once established in a woodland/scrub/hedgerow, boxes can be used in the same basic way as tubes. A minimum of 50 boxes (30 per ha) are installed as a grid 20 m apart, for long-term density estimates, while a line of boxes 20 m apart on a woodland edge may achieve higher occupancy rates. The same detectability probability score is used by both boxes and tubes. Sampling is done monthly or bimonthly. Boxes are used throughout the year, but with a peak in May (Chanin and Gubert 2011).

(continued)

Table 5.4 Species guidance: Hazel Dormouse (*continued*)

Species: Hazel Dormouse	
	3. Hair tubes. These are plastic tubes, fixed to the underside of branches, baited internally with jam or peanut butter, and use sticky tape by the opening. As dormice enter to access the food, their hair sticks to the tape. With a success rate of around 10%, the proof of presence is hard to achieve, although they may work within a week or two of installation, far faster than boxes or tubes. Failure to provide hairs is not the same as absence. 4. Nests. Natural nests are best detected in late autumn/winter after leaf fall and before destruction by the elements, use as bird nest material or being obscured by spring regrowth. Failure to find them is not proof of absence. 5. Trapping. Due to the normal density of dormice, use of licenced baited traps is a relatively hit and miss process and needs a licence. It is not recommended (Bright et al. 2006) and is resource-intensive, as daily visits are required. If boxes and tubes are put up speculatively, no licence is needed. Once a dormouse is found, all later inspections need to be licensed.
Survey timing	As dormice hibernate from October to April, active survey time is between these months. For signs of nut use, mid-autumn is best. For basic habitat assessment – mindful of the breadth of habitats used – this can be conducted year-round.
Timing of visits in the day	If nut or nest surveys are undertaken, then this is dependent on time available, rather than fitting any behavioural time in the day. Out of hibernation, activity peaks are between dusk and dawn, which in turn are affected by temperature and weather (Juskaitis and Buchner 2013). Tube inspections during the day will pick up dormice in their nests.
Visit frequency	If checking tubes or boxes, the recommended frequency is monthly, whilst hair tubes may require more frequent visits. Trapping requires daily visits.
Survey duration	For a full tube/box survey, the recommended survey period is April to November.
Weather	Weather conditions (wind, rain, temperature) will all affect activity/emergence significantly. In cold early spring periods dormice may go into torpor if in boxes.
Light	Light is a key factor affecting behaviours as dormice are normally nocturnal (Juskaitis and Buchner 2013).

SURVEYS FOR PROTECTED SPECIES: WHAT THE LPA MIGHT HAVE ORDERED | 109

Species: Hazel Dormouse	
Distance	Searches up to 1 km from the site boundary will be needed to indicate the presence of other populations to those that may be present in addition to populations on the development site. Individual movements in dispersing young vary across the season, with those born in May having a mean dispersal distance of 363 m and a maximum of 1,300 m. Later born dispersal is around 127 m, confirming the need to check adjacent areas (Juskaitis and Buchner 2013).
How far beyond development boundaries	As dormice occur at low density, and do move up to 1 km on dispersal, 1 km beyond the site limit is needed.
Years of data	In areas with hazel, at least one year of sampling should be undertaken. To establish status in areas without hazel requires use of boxes and tubes and may require up to several years (Bright et al. 2006). Bright et al. (2006) note that due to seasonal issues, survey work may last for more than one year.
Limitations	Dormouse activity is influenced by time of year, weather, temperature, availability of food sources and habitats. For nut sampling, hazel are required and need checking in early autumn. To establish use of an area requires at least 50 tubes/boxes at a standard spacing and repeat visits over the April to November season.
Suitable licensed and experienced surveyors	Surveyors are expected to have practical experience (CIEEM 2013b), under the supervision of a licensee, and to have: A) attended appropriate training courses or received an equivalent level of training from an existing licensee B) field work experience, demonstrated by a logbook of visits on at least 10 separate occasions to check nest boxes or tubes. This should include visits to at least three different sites (to demonstrate the ecologist understands the variation in habitats used by dormice); and C) handling a minimum of five dormice in varying life stages (if planning to handle dormice). Where disturbance is expected, a licence must be held.
Determining impacts	Assessing impacts will require knowing the locations of nests, areas used and preferred habitats. Negative impacts include habitat loss, breaking of linear connections between areas and increased levels of post-development habitat disturbance. Winter disturbance may well be a significant impact on hibernating dormice (Bright et al. 2006). Fragmenting habitats is negative, and any planned compensation areas will need to be planted and allowed to grow for several years in advance. The simplest option is to avoid disturbing known populations.

(continued)

Table 5.4 Species guidance: Hazel Dormouse (*continued*)

Species: Hazel Dormouse	
Mitigating properly assessed potential impacts	Basic principles should include: development that avoids the most heavily used habitats; identifying suitable areas for mitigation by developing similar habitats but with access and low disturbance likely. Any habitat compensation will need to be well established before use by dormice and include reconnection of links to other areas or new links in fragmented areas (Bright et al. 2006). This will focus on understorey shrub communities which offer the best food options for dormice. Box provision may help provide nest sites on shrubs.
Licensable activities	Disturbing dormice, or areas known to be used by dormice, even if not there at the time, requires a licence. Any habitat alteration will require a hazel dormouse mitigation licence issued by Natural England.
Monitoring	If mitigation is planned, then it is normally a requirement for it to be monitored. Objectives will need to be stated. Timing will meet normal survey methods and frequency, and require several periods of observation per year. A minimum of a well-budgeted two years of post-development monitoring will be needed, with up to 10 years in the most complex cases (PBP: SGN 2020) as dormouse populations fluctuate year-to-year. To gauge population status a box-based regime will be best.
Summary	A PEA is required, with a searches up to 1 km beyond boundaries. Detection is influenced by time of year: best April to November. A range of signs are used to determine status. For status assessment, box/tube monitoring from April to November is required. Mitigation needs to assess habitat use, limit isolation and fragmentation and requires 2 or more years of post- licence monitoring to assess effectiveness.

PEA = Preliminary Ecological Appraisal; NE = Natural England

Table 5.5 Species guidance: Freshwater Pearl Mussel

Species: Freshwater Pearl Mussel	
Site name and status	FPM occur in designated and non-designated rivers. In England, site status can be gleaned from Natural England's MAGIC website and is a basic part of a PEA desk search. The walkover element of a PEA will indicate suitable habitat and possible evidence of occurrence. Protected under the Wildlife and Countryside Act (1981) and Nature Conservation (Scotland) Act 2004, it should also be listed in the application.
Zone of influence	FPM occur in a limited number of fast-flowing non-calcareous streams and rivers and are affected by changes in water volume and quality, both of which can affect the condition of the riverbed gravel that they rely on as well as their early-stage salmonid hosts. Any proposed action on a riverbed requires surveys at least 100 m upstream and 500 m downstream (NS 2020a). As FPM are also affected by runoff (sediment, agrichemicals, fertilisers, nutrient-rich runoff, especially soluble reactive phosphorus, to which FPM have a tolerance threshold of 0.03 mg/l(NS 2020a)), potential wider impacts need to be considered if there will be activities further upstream that will affect FPM. NS (2020) asks for consideration in all areas of FPM Special Areas of Conservation catchments, rather than immediately nearby. This means that potentially deleterious activities may be much further than the riverbed distances cited in NS (2020) and may include significant parts of a drainage catchment (NS 2003, 2020; CIEEM 2020).
Identified in pre-application discussions	Pre-application discussions will normally include all aspects of biodiversity, especially protected species. Most LPA discussions will refer to validation checks and these will include FPM. If they are not in the discussion, this should be queried by LPAs.
The species should be picked up in the PEA	PEA will include desk survey data sought from Local Record Centres and specialist groups, as well as field surveys. PEA distances extend up to 1 km beyond site boundaries. Absence from a desk search is not the same as site absence (Natural England 2022a). CIEEM (2020) noted that extensive surveys had revealed new populations of FPM, confirming under-recording in many locations. CIEEM also noted that stream/river size supporting FPM varied significantly, so that stream/river size alone was not a predictor of possible status and scoping out is needed at an early stage of considering possible developments. Extended phase 1 surveys should pick up potential FPM habitat: fast-flowing, non-calcareous streams and rivers with trout or salmon populations and river gravels. FPM can only be detected in situ by risking disturbance, so a licence is required for surveying in the water.

(continued)

Table 5.5 Species guidance: Freshwater Pearl Mussel (*continued*)

Species: Freshwater Pearl Mussel	
Protected species in and around the site	After using the PEA desk search as a screening for records in and around the site, walked coverage should extend 1 km beyond site boundaries to evaluate the potential for recolonisation if the site is potentially impacted or habitats lost.
Age of data attached to the application	Field data are out of date within two years (CIEEM 2019b; Natural England 2022a).
Objectives of surveys	For FPM, the objective of the walkover survey in the PEA is to provide outline evidence of suitable habitat presence. Establishing presence requires formal surveys by two surveyors (CIEEM 2020; NS 2020a). The walkover survey will also produce a map of phase 1 habitats (JNCC 2010), which will provide an indication of the main habitat types in and around the site and the existing local sources of potential nutrient and silt inputs. The main objective of a detailed phase 2 survey is to establish the presence of FPM, the areas supporting FPM and an indication of age and abundance. Survey methods should be explicitly stated so that they can form the basis for any post-development monitoring. If the potential activity involves a local point-source development, then the 100 m above and 500 m below survey area may suffice. If the effect will be to alter water quality or quantity over the wider catchment, then much larger areas will need covering. Context is critical.
Field collection methods	For health and safety reasons, two surveyors are needed. Depending on the potential development, the following methods should be used. If there are potentially direct interventions, then the entire riverbed in the affected areas will need to be sampled using a 1 × 1 m grid. Surveyors enter into the water, and work upstream – using a glass-bottomed viewing bucket – in water deep enough for safe wading. The numbers of mussels are counted and measured in each grid square. Searches for juvenile and hidden mussels should also be carried out in 20% of the square in which visible mussels are recorded. If the effect would be to reduce flows, then the entire area would need checking. This is done by identifying potentially suitable sections of riverbed and using the same basic methods as the direct impact method: 1. work upstream, moving loose debris or weed aside, but without disturbing the riverbed, or 2. check under cobbles, boulders or overhanging banks. 3. If no mussels are found, move to other areas – recording equally for those areas with and without mussels on a standard recording form. This includes details of substrate, algal cover, bankside vegetation and flow characteristics (Young et al. 2003).

Species: Freshwater Pearl Mussel	
	If mussels are found, then search a 50 × 1 m transect traversing the habitat. If <250 mussels are found, then all areas are covered. If there are more, then 1 m quadrats at 10, 20, 30, 40 and 50 m intervals should be searched. Emphasis is on size and potential presence of young mussels. Abundance levels are ranked from 0 to 1,000 individuals. NB: trials of eDNA have not proved effective in determining presence/absence of FPM (CIEEM 2020).
Survey timing	Fieldwork is only possible in periods of low water: April to September. This is due to water levels and also to low turbidity, allowing clear visibility. Visits after heavy rain are unsuitable.
Timing of visits in the day	As fieldwork depends on clear visibility, surveys should be undertaken in maximal hours of day light, rather than in poor light, normally 10.00 to 16.00 and bright sunshine (Young et al. 2003).
Visit frequency	FPM surveys are normally a single visit (Young et al. 2003; NS 2020a).
Survey duration	Undertaken once between April and the end of September, survey duration depends on light levels – which may require later visits if poor.
Weather	In unsuitable weather conditions (cold and high water), mussels may become less visible and affect detection.
Light	Light is a key factor affecting detection.
Distance	For mechanical impacts, at least 100 m above the site and 500 m below. Where water volume and quality are likely to be affected, then a much larger watershed coverage is needed (Young et al. 20023; NS 2020a).
How far beyond development boundaries	See distance above.
Years of data	A properly carried out survey will take a number of days. Given the longevity of the FPM and the recruitment into older age bands, a repeat every 6–10 years is needed (Young et al. 2003; NS 2020a).
Limitations	Surveys are affected by water levels, water clarity, temperature, turbidity and light. As salmonids form a life stage of the FPM, knowing presence/absence of possible host species is required, ideally at least 0.2 juvenile fish per m^2 (Young et al. 2003).
Suitable licensed and experienced surveyors	Licensed surveyors are expected to have previous practical experience (Young et al. 2003; NS 2020a), but there are no CIEEM competencies to judge against.

(continued)

Table 5.5 Species guidance: Freshwater Pearl Mussel (*continued*)

Species: Freshwater Pearl Mussel	
Determining impacts	Assessing impacts will require knowing the potential scale and form of impact. Negative impacts will include direct habitat loss and more diffuse impacts through water quality and volume changes that may well alter depth of water and the necessary substrate used by mussels, as well as affecting the fish host populations. Altering these risks, limits the recruitment of individuals into the larger reproductive stages of life.
Mitigating properly assessed potential impacts	Avoidance of any mechanical impact or indirect impact would be the very basic requirements for FPM retention. Because of the protected status of FPM, a site management plan is required to demonstrate objectives, impacts, mitigation and then post-development monitoring. If works are needed, then precautions are needed to stop silt or sediment entering into water courses, and these themselves need controlling to avoid wash into the river. Special Areas of Conservation prescribe watercourse buffer strips, avoidance of fish spawning and hatching periods (October to May) and avoidance of water course crossings. Mitigation would also involve not changing water levels and nutrient status within the catchment that might impact FPM habitat conditions. Natural England (2022d) suggests that compensation actions should include: improving habitat, reconnecting habitats, potentially undertaking captive breeding and release, as well as providing better conditions for host salmonids. These actions should be part of an approved site management plan.
Licensable activities	Disturbing FPM requires a licence. Any habitat alteration will require a licence. Natural England will only issue a license if the proposal enhances FPM habitats and helps to conserve them.
Monitoring	If mitigation is planned, then it normally requires monitoring. Objectives will need to be stated. Timing will meet normal survey methods and frequency. To gauge population status, a 6–10-year return time is needed (Young et al. 2003).
Summary	A PEA is required, with variable search distances according to potential agent of impact. Surveys are undertaken April to the end of September and detection is influenced by light, water levels, turbidity, temperature and water depth. Mitigation needs to assess habitat impacts, and if unavoidable, includes sediment, water level change and nutrient status controls and habitat protection. Monitoring is based on assessing age range and reproduction in cohorts every 6–10 years.

FPM = Freshwater Pearl Mussels; PEA = Preliminary Ecological Appraisal; NE = Natural England; NS = NatureScot; CIEEM = Chartered Institute of Ecological and Environmental Management

Table 5.6 Species guidance: Water Vole

Species: Water Vole	
Site name and status	WV occur in designated and non-designated wetland habitats and locally away from wetland margins (Strachan et al. 2011; Dean et al. 2016; Dean 2021). In England, site status can be gleaned from Natural England's MAGIC website and is a basic part of a PEA desk search.
Zone of influence	WV occur in a wide range of environments, from fast flowing upland and lowland streams to canals and broad watercourses. Normally (Strachan et al. 2011; Dean 2021), they are found within 5 m of the water's edge. Territories range up to 350 m (Strachan et al. 2011), so searches need to consider far larger linear distances than from waterbody margins. In England, terrestrial populations occur near Newcastle and on the Humber estuary (Dean et al. 2016). The ZoI to check depends on the potential scale of any development (Dean et al. 2016), with temporary local, small-scale, impacts (<15 m) requiring checks 100 m upstream and downstream beyond the footprint, with desk checks and field checks for alternative habitats 1–2 km upstream and downstream. The ZoI increases with length impacted. For temporary impacts, >50 m status checks should be 500 m upstream and downstream, with desk checks and habitat checks looking at 2–5 km. For permanent changes >50 m the ZoI, status checks should be 250 to 500 m upstream and downstream, with desk and spot checks up to 2 km. For large scale permanent coastal realignment works, the field checks should be up to 1,000 m and the desk checks up to 10 km.
Identified in pre-application discussions	Pre-application discussions will normally include all aspects of biodiversity, especially protected species. Most LPA discussions will refer to validation checks and these will include WV. If they are not in the discussion, this should be queried by LPAs. Protected under the Wildlife and Countryside Act (1981,) and listed as rare and most threatened species under section 41 of the NERC Act 2006. WV should be listed in the application.
The species should be picked up in PEA	The PEA will include desk survey data sought from Local Record Centres and specialist groups, as well as field surveys. The walkover element of a PEA will indicate suitable habitat and possible evidence of occurrence. PEA distances extend up to 10 km beyond site boundaries, depending on nature and scale of potential impact (Dean et al. 2016). Absence from a desk search is not the same as site absence (Natural England 2022a).

(continued)

Table 5.6 Species guidance: Water Vole (*continued*)

Species: Water Vole	
	As WV populations have reduced markedly since 2000, NatureScot (NS 2020a) cautions against using record centre data that pre-date 2000 as background. Dean et al. (2016) also caution against use of data without collection contexts. Field level data should be no more than two years old (NS 2020a; CIEEM 2019a). Dean et al. (2016) noted that an extended phase 1 walkover as part of a PEA could be used to assess the presence of preferred habitats: especially the close proximity of dry areas above water level for nesting, herbaceous vegetation for nesting or feeding and water as a means of predator escape. Dean (2021) indicated the strong differences according to season in habitats and noted the highly seasonal (April–October) pattern of WV occurrence and detection probability, and the need for two visits (Strachan et al. 2011). This means that using a single PEA visit as categorical for presence/absence of WV is unsafe.
Protected species in and around the site	After using the PEA desk search as a screening for records in and around the site, walked coverage should extend beyond site boundaries according to the scale and potential permanency of the proposed development to evaluate the potential for recolonisation if the site is potentially impacted or habitats lost (see ZoI).
Age of data attached to the application	Field data are out of date within two years (CIEEM 2019b; Natural England 2022a, NS 2020a).
Objectives of surveys	For WV, the objective of the walkover survey in the PEA is to provide outline evidence of suitable habitat presence, and if in the correct season some possible evidence of vole presence. Establishing presence requires formal surveys by experienced surveyors (CIEEM 2020; NS 2020a) and is rarely met by a single PEA visit. The main objective of a detailed phase 2 survey is to establish the presence of WV and an indication of abundance. Survey methods should be explicitly stated so that they can form the basis for any post-development monitoring.
Field collection methods	Depending on the potential scale and permanency of development (see ZoI), the following methods should be used. For distance see ZoI. Water Vole surveys include both habitat and field sign surveys, ideally including actual sightings. As WV habitats alter significantly over the course of the season (April to the end of September, slightly shorter in upland areas (Strachan et al. 2011; Dean et al. 2016; Dean 2021)) and voles may use different areas across the season, all authors advise against single visits, unless the objective is to note presence, based in almost all cases on field signs of latrines, which may be met potentially on a first visit.

SURVEYS FOR PROTECTED SPECIES: WHAT THE LPA MIGHT HAVE ORDERED | 117

Species: Water Vole	
	Two visits – at least two months apart – are recommended: 1. From mid-April (as early as mid-March in SE England). 2. July to the end of September inclusive. Second visits are important as, after breeding, population density will be at its highest and may show areas in use from colonies outside of the area that were absent on the first visit. Signs are critical for determination of WV presence: Signs of WV (Strachan et al. 2011) include: – faeces: the only field sign to be used reliably on its own – latrines: most are found where animals enter and leave the water, at range boundaries, or near the nest – feeding stations: food items are usually taken to a preferred site where they are eaten, or cut up for taking into burrows for later use. Plant species identification can help determine diet – burrows: the 4–8 cm burrows (wider than taller) are usually close to the water's edge, although some may be up to 5 m inland from the water. Burrows will include nest chambers – lawns: grazed areas may be found around land holes – footprints: these may be found on wet mud by the water's edge, and are distinctive in size and shape – runways: these are low tunnels pushed in vegetation and may lead to water, preferred feeding areas or burrow entrances. The number of latrines will indicate both the relative population size, and the most valuable parts of a site for voles: April to June: high ($\geq$10+/100 m of bankside habitat); medium (3–9); low $\leq$2, or none but with other confirmatory field signs July to Sept: high (20$\geq$+/100 m of bankside habitat); medium (6–19); low $\leq$5, or none but with other confirmatory field signs.
Survey timing	In lowland South East England, March to October; upland England mid-May to mid-Sept; otherwise, mid-April to end of September. Visits after heavy rain are unsuitable.
Timing of visits in the day	Within the day, good light conditions are needed to detect for signs.
Visit frequency	Two separate visits at least two months apart.
Survey duration	Undertaken between April and the end of September, survey duration depends on vegetation, water depth and methods used (waders/boats may be needed in some locations).

(continued)

118 | PROTECTED SPECIES AND BIODIVERSITY

Table 5.6 Species guidance: Water Vole (*continued*)

Species: Water Vole	
Weather	Visits should not be after heavy rain or rising water levels when many of the signs of voles will be washed away.
Light	Light may affect detection.
Distance	The minimum distance to sample will require checks 100 m up and downstream beyond the footprint.
How far beyond development boundaries	See distances in ZoI above.
Years of data	Properly carried out surveys require one year of data. Historical data can help, but have context issues. Annual monitoring should be up to five years according to project type and extent. (Dean et al. 2016).
Limitations	Surveys are affected by water levels, time of year, weather, vegetation, management practices, experience (Dean et al. 2016; Dean 2021) and marked vegetation changes over the growing season.
Suitable licensed and experienced surveyors	Because of the risk of disturbance, formal WV surveys require a licence. Licensed surveyors are expected to have previous practical experience and meet the CIEEM Water Vole competencies (CIEEM 2021). Licences are also required to disturb voles or their habitats.
Determining impacts	Assessing impacts will require knowing the potential extent and form of impact and an indication of the scale of the population using part or the whole area. Impacts will include: – habitat loss and deterioration – fragmentation of habitats or closing dispersal through fixed installations such as culverts – incidental mortality – damage to burrows – water regime changes – pollution leading to secondary habitat effects – noise. The timing and scale of impacts will determine the overall effects. NB: as time will have elapsed from initial surveys, pre-construction surveys will be recommended for all areas thought to contain or be likely to contain WV, and may require new mitigation hierarchy activities.
Mitigating properly assessed potential impacts	Avoidance of any mechanical impact or indirect impact would be the very basic requirements for WV retention. Suitable avoidance actions (Dean et al. 2016) would include: – retaining current habitats in situ – putting a 10 m exclusion buffer around WV habitat – avoiding new culverts or using oversize culverts – use of no dig/directional drilling to avoid habitat disturbance.

Species: Water Vole	
	If works are going to be bigger than 50 m of bank, Natural England would need to issue a licence to transfer WV to a pre-prepared habitat area (established at least one year ahead (Dean et al. 2016)) suitable for the translocation of a population. For Nature Scotland, this is an exceptional action, due pre-breeding season (NS 2020a): 1 March to 15 April (Dean et al. 2016), but not thereafter. Dean et al. (2016) are uncertain of the proven utility of translocation.

Compensation measures for WV include:
– extending and improving suitable habitats for WV without fragmenting existing habitats
– improving water quality.

Because of the protected status of WV, a site management plan will be required to demonstrate objectives, impacts, mitigation and then post-development monitoring. This should cover 3–5 years and be funded to include both the survey element and also any remedial works that arise out of the monitoring (Natural England 2022e; NS 2020a). |
| Licensable activities | Disturbing WV requires a licence. Any habitat alteration will require a licence. Natural England will only issue a license if the proposal enhances WV habitats and helps to conserve them (Natural England 2022e). |
| Monitoring | If mitigation is planned, then it normally requires monitoring. Objectives will need to be stated. Timing will meet normal survey methods and frequency. To gauge population status, a 3–5 year monitoring period is needed (Dean et al. 2016; 2021). |
| Summary | A PEA is required, with variable search distances according to potential agent of impact. Two surveys, two months apart, are undertaken April to end of September and detection is influenced by light, water levels, vegetation condition and management. Of the detection signs, droppings at latrines are the definitive proof.
Mitigation needs to assess habitat impacts, and if unavoidable, will require site management, and could potentially include relocations. Efficacy of these is uncertain. |

WV = Water Voles; PEA = Preliminary Ecological Appraisal; NE = Natural England; NS = NatureScot; CIEEM = Chartered Institute of Ecological and Environmental Management; ZoI = zone of influence

Table 5.7 Species guidance: Otter

Species: Otter	
Site name and status	Otters occur in designated and non-designated wetland habitats and locally away from wetland margins. In England, site status can be gleaned from Natural England's MAGIC website and is a basic part of a PEA desk search.
ZoI	Otters occur in a wide range of environments, from fast flowing rivers and lowland streams to canals and still waterbodies and coasts (Crawley et al. 2020). Individual home ranges may be up to 35 km long, and individuals may move several kilometres overnight (Chanin 2003b; Natural England 2020a). This means that only large catchments can be expected to be self-sustaining (Chanin 2003b). Sampling smaller lengths will only pick up elements of a population. Using information from national Otter surveys and Local Record Centres will help understand distributions within the site and wider area (CIEEM 2020) and the potential wider ZoI. The effective ZoI near a proposed development is larger than the immediate 200 m survey minimum noted by Nature Scotland (2020; CIEEM 2020).
Identified in pre-application discussions	Pre-application discussions will normally include all aspects of biodiversity, especially protected species. Most LPA discussions will refer to validation checks and these will include Otters. If they are not in the discussion, and they are known from national surveys to have been in the area, this should be queried by LPAs.
The species should be picked up in the PEA	The PEA will include desk survey data sought from Local Record Centres and specialist groups, including mammal groups, as well as field surveys. The walkover element of a PEA will indicate suitable habitat and possible evidence of occurrence. This is primarily the occurrence of spraints near bridges and weirs or footprints in mud/soft substrate (Chanin 2003b; Natural England 2022f). PEA distances extend at least 200 m (NS 2020a) or 250 m (Natural England 2022f) beyond proposed site boundaries, and 50 m on either side of the waterbody (Natural England 2022f). CIEEM (2020) notes that for many projects several survey visits throughout the year may be required, even where a PEA habitat and signs of Otter during the field visit may suggest Otter presence. Nature Scotland (NS 2020a) cautions against using record centre data that are older than two years. Field-level data should be no more than two years old (NS 2020a, CIEEM 2019). Absence from a desk search or lack of initial site evidence is not the same as site absence (Natural England 2022a; Natural England 2022f; PBP 2021). CIEEM (2020) and Chanin (2003b) indicated the strong seasonal differences in detection, as vegetation alters visibility within habitats across summer months.

SURVEYS FOR PROTECTED SPECIES: WHAT THE LPA MIGHT HAVE ORDERED | 121

Species: Otter	
Protected Species (PS) in and around the site	After using the PEA desk search as a screening for records in and around the site, walked coverage should extend beyond site boundaries according to the scale and potential permanency of the proposed development, with at least 200 m beyond boundaries and being 50 m wide along both sides of water courses.
Age of data attached to the application	Field data are out of date within two years (CIEEM 2019; Natural England 2022a; Natural England 2022f; NS 2020a).
Objectives of surveys	The objective of the walkover survey for Otters in the PEA is to provide outline evidence of suitable habitat (JNCC 2010) and possible evidence of Otter presence, mainly through spraints (Chanin 2003a,b). The main objective of a detailed phase 2 survey is to establish the presence of Otters in the proposed and wider areas. Although spraints indicate presence, their number/volume is not a simple indicator of Otter numbers (Chanin 2003b; Balestrieri et al. 2011). Establishing presence requires formal surveys by experienced surveyors (CIEEM 2020; NS 2020a; Chanin 2003b) and is rarely met by a single PEA visit or detection visit. Balestrieri et al. (2011) indicated strong inter-year variability in Otter detection and a minimum requirement of three visits in a year in order to establish Otter presence. The extent of survey detail varies with the survey objective. If the objective is to cover large areas and limit time, then there is a focus on spraint detection at bridges and similar features and areas 50 m upstream and downstream (Chanin 2003b, 2005). More detailed assessments in a proposal area will require more time and recording: see methods. Survey methods should be explicitly stated so that they can form the basis for any post-development monitoring.
Field collection methods	Diurnal Otter surveys include both habitat and field sign surveys; the normally nocturnal individuals are rarely seen during surveys. To produce a monitoring baseline, sites should be surveyed annually for five years, and then checked at three-yearly intervals. Chanin (2003b) describes the main methods: – presence of spraints under bridges or at weirs: this is the 'standard' method to determine presence over wide areas and can also be used in and near development sites. Checks should not be made after heavy rain, and are best five or more days after rain. They are best between May and the end of September. Spraints may also occur near holts/resting areas. Chanin noted that the associated habitat recording (Chanin 2003b) is potentially time consuming.

(continued)

Table 5.7 Species guidance: Otter (*continued*)

Species: Otter	
	– presence of tracks on damp substrate: affected by high water flows and heavy rain. A possible alternative to the standard method where bridges and similar are not available in the area. Print size will indicate sex (male: large; female: medium; small: juveniles) (Chanin 2003b). Tracks can be seen in snow. – holt/den presence. On coastal areas, dens are limited, so can be searched for. In riverine locations with long territory lengths, and many possible features for den use, this is a far less reliable indicator (Chanin 2003a). As there may be many ephemeral dens along a big territory, absence of use may not be a categorical sign of absence. Natal holts are used for birth and raising young and will be associated with well-used areas. – road kills. As a proxy for a population, road kills indicate the presence of Otters, but are not directly correlated with population levels (Chanin 2003a,b). – food remains: Otters commonly leave partly eaten large fish, especially heads, within their territory, and these may be used as indicators of presence when few other signs are present. As vegetation grows, detection of these becomes more difficult. – slides: these are sloping areas where Otters repeatedly enter water, and may be worn down by use; they regrow quickly if unused in the growing season. – couches/resting places: these are grassy areas used for resting and may include dry grass pulled in from nearby. – DNA records. Spraints can be tested for confirmation, and also to identify individuals, but this is an expensive way of confirming status. – Camera traps: where Otters are thought to be about, camera traps can be used if they will not disturb Otters. If there is a risk of this, then a licence is needed (Natural England 2022f). Torches or endoscope use also need a licence. Surveys are best undertaken May to the end of September when water levels are lower, chances of spraint being washed away are reduced, and rainfall, which affects spraints and other indicators, is also less frequent. Surveys can be undertaken from the bank, or in waders in areas of low water. Both banks can be worked in pairs, or from a boat, but this risks missing many land-based features. Habitat maps of the 50 m bands either side of rivers record the main habitats, presence of potential holts/dens and water levels.
Survey timing	Fieldwork is possible all year, but best in periods of low water: April to September. This is due to water levels; visits after heavy rain are unsuitable. Areas such as sand bars are more exposed in low water.

SURVEYS FOR PROTECTED SPECIES: WHAT THE LPA MIGHT HAVE ORDERED | 123

Species: Otter	
Timing of visits in the day	As Otters are mainly nocturnal, most sign-based surveys can be undertaken in the day. If direct Otter presence is suspected in a den/holt, care must be taken not to disturb the location unduly. Camera traps will operate best at night and should have silent running and avoid using monitoring light indicators.
Visit frequency	For effective status assessments, several repeat surveys are needed. Where breeding/resting sites may be affected, several months of camera traps are required (NS 2020a).
Survey duration	Survey duration depends on vegetation, water depth and methods used (waders/boats may be needed in some locations) and varying water levels – which may require later visits if poor. If monitoring of standard spraint sites is used, then this depends on distances between features such as bridges.
Weather	Visits should not be after heavy rain or rising water levels when many of the signs of Otters will be washed away. Surveys are best five days or more after rain.
Light	Light is not an obvious limiting factor in most sign detection.
Distance	The minimum distance to sample will require checks 200 to 250 m upstream and downstream beyond the site footprint. If the objective is to establish the presence of Otters beyond the site and across the larger immediate area of a drainage system, this would require systematic inspection of bridges and similar spraint sites within >5 km (Chanin 2003b; Natural England 2022f)
How far beyond development boundaries	A minimum of 200–250 m (Natural England 2022f; NS 2020a).
Years of data	At least one year's worth of several visits. Chanin (2003b) recommends annual surveys for five years to fully establish a baseline. Where Otters are known to be close to a potential development site, pre-construction surveys should occur within three months or fewer of development commencement (NS 2020a).
Limitations	Surveys are affected by water levels, time of year, weather, vegetation, management practices, experience and marked vegetation changes over the growing season.
Suitable licensed and experienced surveyors	Because of the risk of disturbance, some aspects of Otter surveys require a licence. Licensed surveyors are expected to have previous practical experience and meet the CIEEM Otter competencies (CIEEM 2013f).

(continued)

Table 5.7 Species guidance: Otter (*continued*)

Species: Otter	
Determining impacts	Assessing impacts will require knowing the potential extent and form of impact and an indication of the scale of the population using part or the whole area. Impacts will include: – habitat loss and deterioration – fragmentation of habitats or closing dispersal through fixed installations such as culverts – incidental mortality – damage to dens/holts – water regime changes – pollution leading to secondary habitat effects/fish effects and carrying capacity – noise. The timing and scale of impacts will determine the overall effects. NB: as time will have elapsed from initial surveys, pre-construction surveys will be recommended for all areas thought to contain or be likely to contain Otters and take place <3 months ahead of planned commencement (NS 2020a).
Mitigating properly assessed potential impacts	Avoidance of any mechanical impact or indirect impact would be the very basic requirement for Otter retention. Suitable avoidance actions (NS 2020a) would include: – retaining current habitats in situ – putting a 30 m exclusion buffer near non-breeding holts, >200 m where breeding is occurring, unless naturally shielded from disturbance – building new tunnels/culverts to preclude pushing Otters onto road in high water – avoidance of night-time working or work within two hours of dawn and dusk – use of Otter-suitable fencing to stop Otters entering the site. Mitigation measures for Otters will include: – extending and improving suitable habitats, without fragmenting existing habitats – improving water quality – building a new holt/den (under licence) if old holt has to be destroyed – improve habitats/habitat connectivity. Because of the protected status of Otters, a site management and monitoring plan will be required to demonstrate objectives, impacts, mitigation and then post-development monitoring. This should cover three to five years and be funded to include both the survey element and also any remedial works that arise out of the monitoring (Natural England 2022f; NS 2020a).

Species: Otter	
Licensable activities	Disturbing Otters requires a licence. Any habitat alteration will require a licence. Natural England will only issue a license if the proposal enhances Otter habitats and helps to conserve them.
Monitoring	If mitigation is planned, then it normally requires monitoring. Objectives will need to be stated. Timing will meet normal survey methods and frequency. To gauge population status, a five-year baseline monitoring period is needed (Chanin 2003b, 2005).
Summary	A PEA is required, with variable search distances according to the potential agent of impact. Several surveys, two months apart, are undertaken from April to the end of September, and detection is influenced by water levels, vegetation condition and management. Surveys should be at least 200 m beyond proposed site boundaries. Wider coverage is needed to fully understand potential population impacts. Mitigation needs to assess habitat impacts and, if unavoidable, will require site management and could potentially include holt removal and replacement. The efficacy of mitigation is unquantified.

PEA = Preliminary Ecological Appraisal; ZoI = zone of influence; NE = Natural England; NS = NatureScot; CIEEM = Chartered Institute of Ecological and Environmental Management

Table 5.8 Species guidance: reptiles

Species: reptiles	
Site name and status	Reptiles occur in designated and non-designated habitats across the UK. In England, site status can be gleaned from Natural England's MAGIC website and is a basic part of a PEA desk search.
ZoI	In England, reptiles occur in a wide range of habitats, and as a result there is no simple distance prescription to cover all species. Dispersal distances vary between species and across the year. Adult male Sand Lizards in search of females may move 500 m, and dispersing young may move similar distances (Edgar and Bird 2006). Forestry Commission et al. (2013) give a 100 m low risk threshold for Sand Lizard and Smooth Snake disturbance. Francois et al. (2021) recommend a 500 m buffer zone for Adders due to male dispersal, and Grass Snakes (Edgar et al. 2010) move up to several kilometres in an active season, so a ZoI that reflects this would be around 500 m. Common Lizards and Slow Worms, in contrast, would have a ZoI of 100 m (Edgar et al. 2010).

(continued)

Table 5.8 Species guidance: reptiles (*continued*)

Species: reptiles	
Identified in pre-application discussions	Pre-application discussions will normally include all aspects of biodiversity, especially protected species. Most LPA discussions will refer to validation checks, and these will include reptiles in a number of areas. If they are not in the discussion, this should be queried by LPAs.
The species should be picked up in the PEA	PEA will include desk survey data sought from Local Record Centres and specialist groups, reptile specialist groups, as well as field surveys. The walkover element of a PEA will indicate suitable habitat and possible evidence of occurrence. SGN (2021) and Edgar et al. (2010) both noted that absence in brief searches during a PEA visit would not be definitive due to the levels of effort, timing issues and weather issues related to detailed searches. Given the ZoI distances of 100 m for all but Adders (500 m), Sand Lizards (500 m) and Grass Snakes (500 m), a search zone of 100 m beyond the site boundary would suffice. Neither NE nor NS advise on distances. Both Natural England (2022g) and Nature Scotland (NS 2020a) caution against using record centre data that are older than two years. Field level data should be no more than two years old (NS 2020a, CIEEM 2019b). Absence from a desk search or lack of initial site evidence is not the same as site absence (Natural England (2022g; PBP 2021)).
Protected species in and around the site	After using the PEA desk search as a screening for records in and around the site, walked coverage should extend beyond site boundaries as noted in the ZoI.
Age of data attached to the application	Field data are out of date within two years (CIEEM 2019b; Natural England 2022g; NS 2020a).
Objectives of surveys	The objective of the survey will determine the methods used and the level of effort, which will also be affected by the time of year and climatic variables (Edgar et al. 2010; Sewell et al. 2013; Froglife 1999; Gent and Gibson 2003). Objectives will include: – presence/absence – enough to stimulate a detailed survey – assessing which areas on a site are used by reptiles – assessing species population sizes – assessing potential habitat loss and management effects – effects of removing populations in case of partial/total area destruction. Subsequent to any development, a round of monitoring surveys would be expected, with these matching techniques pre-development.

Species: reptiles	
Field collection methods	There are two basic methods used in the UK (Edgar et al. 2010; Sewell et al. 2013; Froglife 1999; Gent and Gibson 2003): – directed visual transects – refugia searching (natural and artificial). Directed visual transects are based on looking for animals in sunny, open, well-drained and often south-facing areas previously noted on a site inspection visit. Walked slowly, reptiles may be seen or heard disappearing, in which case, a return to the area within 10–15 minutes is needed. To establish presence requires at least seven visits in suitable weather, at suitable times of the year. To establish which are key areas, or an idea of relative populations sizes, at least 20 visits per season are needed. The exact number of visits varies to some extent with species: a small population of Smooth Snakes may take tens of visits over months to detect (Edgar et al. 2010), while Adders and Grass Snakes use different patches across the season, so to understand their use of a site requires visits spread across the active season. Smooth Snakes and Slow Worms are almost always found under refugia, so any transect asserting their absence would be in error based on that alone (Edgar et al. 2010). Spring visual searches are good for detecting communal hibernation sites. Refugia searching: this uses examination under natural and artificial refuges (iron/felt/onduline (Edgar et al. 2010)). Tin sheeting is widely used (>0.5 m^2) and once 'bedded in' for a few weeks in a sunny but non-obvious location may become used; this depends in part on population density (Edgar et al. 2010). Skin sloughing: occasionally sloughed skins will be found, but given their infrequency, and patchiness this is not a reliable indicator of absence.
Survey timing	Fieldwork is possible for only a limited part of the year, March to October, but desk assessments and first habitat visits can be done in preparation earlier (Gent and Gibson 2003). April, May and September are the key months, with the first two being the breeding season when animals are more visible. In late March, animals emerging early from hibernation are often seen basking. As animals are warmer in June, July and August they are more active and deeper in vegetation, so are less visible (Gent and Gibson 2003). Individual species have slightly varying best times: – Common Lizards are usually visible above 9°C. Unlike others, gravid females continue to bask in June and July. And September is a good time to see young. – Sand Lizards bask in April to May. After breeding in May, males become hard to find. Young hatch in August and are visible for 1–2 months.

(continued)

Table 5.8 Species guidance: reptiles (*continued*)

Species: reptiles	
	– Slow Worms are normally hidden, but may occasionally bask in mid-summer. – Grass Snakes emerge from hibernacula in April, and sites noted then should be visited in September and October as snakes return to hibernate. Snakes bask 12–20°C in obscured locations, with young emerging in September. – Adders emerge early in spring, often basking near hibernacula, with March being the first period for detection. In warm months they spend time basking, often out of sight. Young are seen from August onwards. – Smooth Snakes emerge from April onwards, and do limited basking, but do frequently use refugia; disturbing here requires a licence.
Timing of visits in the day	As reptiles are more active in warm conditions, time of day has a significant effect. Best times are 09.00 to 11.00 and then 16.00 to 19.00.
Visit frequency	For effective status assessments, a minimum of seven visits is needed, with 20 or more spread across the season to fully determine status.
Survey duration	Survey duration depends on the extent of a transect, weather conditions and vegetation and refugia. Cool periods are to be avoided.
Weather	Reptiles are highly affected by weather. Few are out in windy or rainy weather, and in hot periods, basking animals need little time to warm up. Ideal temperatures for most species are 10–17°C with little to no wind. Gravid females bask more than non-gravid females and males.
Light	Reptiles are not active in poor light and cool conditions.
Distance	The minimum distance to sample will require checks at 100 to 500 m depending on species.
How far beyond development boundaries	A minimum of 100 to 500 m.
Years of data	At least one year's worth of visits, with more to establish status for some species, such as Smooth Snakes.
Limitations	Surveys are affected by: time of year, weather (rain, wind, temperature), vegetation growth and height, management practices and surveyor experience.
Suitable licensed and experienced surveyors	Because of the risk of disturbance, smooth snakes and sand lizards require a survey licence. Licensed surveyors are expected to have previous practical experience and meet the CIEEM reptile competencies (CIEEM 2014). Licences are also required to disturb reptiles or their habitats.

SURVEYS FOR PROTECTED SPECIES: WHAT THE LPA MIGHT HAVE ORDERED | 129

Species: reptiles	
Determining impacts	Assessing impacts will require knowing the potential extent and form of impact and an indication of the scale of the population using part or the whole area. Impacts will include: – factors according to individual species ecology (see above) – habitat loss and deterioration – fragmentation of habitats or closing dispersal routes between habitat blocks used in different parts of the season – incidental mortality – water regime changes – pollution leading to secondary habitat effects – noise – risk of fires affecting habitats. The timing and scale of impacts will determine the overall effects. NB: as time will have elapsed from initial surveys, pre-construction surveys will be recommended for all areas thought to contain or be likely to contain reptiles and take place <3 months ahead of planned commencement (NS 2020a).
Mitigating properly assessed potential impacts	Avoidance of any mechanical impact or indirect impact would be the very basic requirements for reptile retention. Suitable avoidance actions (NS 2020a; Natural England 2022g) would include: – retaining current habitats in situ – changing work times to avoid the active season or detected hibernacula – putting a temporary reptile fence to prevent animals going into risky areas – increasing connectivity to suitable habitat areas, with clear evidence of suitability for each species involved – managing areas within the season to make them unsuitable as part of a coherent habitat management package. Worst case scenario: translocation. As part of a coherent set of species plans, translocation may be cited as the only option. This requires clear demonstration that the activity would benefit reptile conservation (Natural England 2022g), but no further guidance is provided by NE.

(continued)

Table 5.8 Species guidance: reptiles (*continued*)

Species: reptiles	
	Any habitat use in translocation should be capable of supporting a viable set of populations in a habitat (probably new and developed ahead of removal (Natural England 2022g)) and should not carry existing populations that would be compromised. Such areas would need to be analogous to the removal areas and at least as large and safe from damage or loss in the long term. It would need to be managed as part of a long-term, funded plan. Translocation has a poor history, and removal of populations from an area is severely affected by weather conditions (Sewell et al. 2013), making for uncertainty in being sure of population depletion. Small sites may take a year, and up to three for large sites (NE 2015). NB: Nash et al. (2020) found no confirmatory evidence for reptile translocation to mitigate for development impacts.
Licensable activities	Disturbing Sand Lizards and Smooth Snakes requires a licence. Any habitat alteration will require a licence. Natural England will only issue a license if the proposal enhances reptile habitats and helps to conserve them.
Monitoring	If mitigation is planned as part of a long-term funded programme, then it normally requires monitoring. Objectives will need to be stated. Timing will meet normal survey methods and frequency, and remediation actions undertaken as a result will be in line with plan requirements.
Summary	A PEA is required, with variable search distances according to potential agent of impact and species. A minimum of 7 visits, spread over the period April- September are required, with 20 plus to establish patterns of habitat use and provide indications of population size. April, May and September are the peak detection months. In addition, surveys should be 100 to 500 m beyond proposed site boundaries. Care needs to be taken to consider individual reptile ecologies in surveys and timing. Wider coverage than the site alone is needed to fully understand potential population impacts. Mitigation needs to assess habitat impacts, and if unavoidable, will require site management. Any translocation requires pre-prepared suitable habitats without pre-existing reptile populations and with long-term security. Mitigation efficacy is uncertain.

PEA = Preliminary Ecological Appraisal; ZoI = zone of influence; NE= Natural England; NS = NatureScot; CIEEM = Chartered Institute of Ecological and Environmental Management

Table 5.9 Species guidance: White-clawed Crayfish

Species: White-clawed Crayfish	
Site name and status	White-clawed Crayfish occur in designated and non-designated habitats across the UK. In England, site status can be gleaned from Natural England's MAGIC website and is a basic part of a PEA desk search.
ZoI	WCC occur in a range of waterbodies, both flowing and standing. Because they are susceptible to upstream-derived and off-site pollution, effects of erosion and siltation, Signal Crayfish and crayfish plague, then it is reasonable to look at a catchment level in the first instance (Peay 2003).
Identified in pre-application discussions	Pre-application discussions will normally include all aspects of biodiversity, especially protected species. Most LPA discussions will refer to validation checks and these will include WCC in a number of areas. If potential WCC sites are involved, and they are not in the discussion, this should be queried by LPAs.
The species should be picked up in the PEA	The PEA will include desk survey data sought from Local Record Centres, National Biodiversity Network and specialist groups as well as field surveys. The walkover element of a PEA will indicate suitable habitat and possible evidence of occurrence. Absence in brief searches during a PEA visit would not be definitive, due to the levels of effort and timing issues, and weather issues related to detailed searches. Natural England (2022h) cautions against using record centre data that are older than two years. Field level data should be no more than two years old (CIEEM 2019b). Absence from a desk search or lack of initial site evidence is not the same as site absence (Natural England 2022h).
Protected species in and around the site	After using the PEA desk search as a screening for records in and around the site, walked coverage should extend beyond site boundaries as noted in the ZoI. Peay (2003) recommends sampling in 500 m blocks. These distances should be used above and below the site boundary.
Age of data attached to the application	Field data are out of date within two years (CIEEM 2019b; Natural England 2022h).
Objectives of surveys	The objective of the survey will determine the methods used and the level of effort, which will also be affected by the time of year and climatic variables (Peay 2003, 2004). Objectives will include: – presence/absence – enough to stimulate a detailed survey, or confirm basic status – assessing which areas on a site are used by WCC – assessing species populations sizes – assessing potential habitat loss and management effects – effects of removing populations in case of partial/total area destruction.

(continued)

Table 5.9 Species guidance: White-clawed Crayfish (*continued*)

Species: White-clawed Crayfish	
	Subsequent to any development, a round of monitoring surveys would be expected, with these matching techniques pre-development. These would be as part of a mitigation plan (Natural England 2022h).
Field collection methods	There are a range of methods. None, as Peay (2003, 2004) observes, are ideal and all, to some degree, produce a biased sample of the true population. Within any area, WCC distribution is patchy in response to habitat use and age profile of WCC (Peay 2004). In addition, Peay (2004) notes that single year samples in monitoring are unsafe and recommends two years in any monitoring process. 1. Standard method: This requires the selection of 5 potentially suitable WCC habitat blocks and a physical search in clear water up to 60 cm deep of 10 potential refuges in each of the 5 500 m blocks. It cannot be done in turbid water, where water is too deep or where the only refuges are in banks. Time taken depends on conditions and WCC density. This produces data on relative abundance, population structure, size distribution and sex ratios. It is best undertaken between July and September. It produces limited habitat disturbance – less than fixed-area sampling. 2. Fixed-area sampling: The number of WCC in five 1×1 m² areas is found by removing all potential refuges from a sample area bottom. This typically takes more time than the standard method, as it requires more samples, but can produce density data, as well as age and size data. Like all other methods, it is undertaken after breeding and moulting between July and September. 3. Trapping (baited): Two visits are needed in >8°C water temperature. The first is to set the baited traps, and the second is to remove them after one night. Depending on the waterbody, they may be set from a bank, in a line across a river, or from a boat in a lake or pond at 5 m distance. Setting traps in favoured habitats will increase take compared with regular spacing. Traps only catch active adults and have low efficiency. Both trapping types require Environment Agency licences. Details are also recorded for habitat locations for each trap as well as background conditions. 4. Trapping (unbaited): Left for 2 or more nights in >8°C water temperature. Samples a wider range of sizes than baited traps, but the efficiency is not known and not readily comparable with baited trapping.

Species: White-clawed Crayfish	
	5. Night viewing: This requires water <1 m deep, clear – not turbid – and moderate to low flows in rivers in water >8°C. As WCC are active at night it provides direct sightings and has a low habitat disturbance effect, but can be used to age animals. The number of individuals in a length is recorded, along with size and habitat details. It is time-intensive, as each session requires a separate night, and there are effects of light. More effective than trapping, and typically takes half the time. Strongly affected by water temperature, flow and clarity. In addition to the data recorded in each method, basic data are needed for: – indicative status of WCC and Signal Crayfish in the unit or catchment – chemical and biological water quality, including calcium concentrations. Care needs to be taken over risk of disease transfer between catchment areas.
Survey timing	The potential time for surveys is May to October inclusive (Peay 2003), but the optimal time is July to September to minimise impacts on WCC.
Timing of visits in the day	For daytime work, maximal light conditions are required, so that late afternoon or shaded light is unsuitable. For night viewing, dark conditions are preferable. For trapping, positioning of traps is best done in good light to allow placement on the bottom of the site.
Visit frequency	For effective status assessments, a minimum of one visit will be enough, but for monitoring purposes a minimum of two years is required due to potential confounding effects in individual years (Peay 2003, 2004).
Survey duration	Three months are possible for surveys, and one visit within this period, subject to confounding issues, will be adequate.
Weather	WCC activity is highly affected by weather: directly by water temperature and indirectly by flow variation and chemical concentration and turbidity.
Light	WCC are active out of daylight, making this an effective period to survey individuals.
Distance	The standard sampling units are 500 m, within which there are typically 100 m subsections.
How far beyond development boundaries	The standard sampling unit is 500 m. Where it is likely that there will be potential effects from an activity upstream, review of much of the potentially affected catchment will be required.

(continued)

Table 5.9 Species guidance: White-clawed Crayfish (*continued*)

Species: White-clawed Crayfish	
Years of data	At least one year's worth of visits to establish status. Peay (2004) recommends two years for proper monitoring data.
Limitations	Surveys are affected by: time of year, weather (rain: water levels, turbidity, water temperature) and surveyor experience.
Suitable licensed and experienced surveyors	Disturbance of WCC requires an Environment Agency survey licence. Licensed surveyors are expected to have previous practical experience and meet the CIEEM WCC competencies (CIEEM 2013h). Licences to alter habitat where WCC are involved requires an NE licence.
Determining impacts	Assessing impacts will require knowing the potential extent and form of impact and an indication of the scale of the population using part or the whole area. Impacts will include: – habitat loss and deterioration – fragmentation of habitats or closing dispersal routes between habitat blocks used by WCC – incidental mortality – water regime changes: including oxygen concentrations – water pollution leading to secondary habitat effects – light pollution. NB: as time will have elapsed from initial surveys, pre-construction surveys will be recommended for all areas thought to contain or likely to contain WCC and take place <3 months ahead of planned commencement (NS 2020a).
Mitigating properly assessed potential impacts	Avoidance of any mechanical impact or indirect impact would be the very basic requirement for WCC retention. Suitable avoidance actions (NS 2020a; NE 2022h) would include: – retaining current habitats in situ – changing work times to avoid the active season – putting temporary fencing to control sediment or another deleterious runoff – reduction of works near banks – temporary fencing to control WCC access to work areas – avoiding introduction of alien crayfish – increasing connectivity to suitable habitat areas, with clear evidence of suitability for each species involved – managing areas within the season to make them unsuitable as part of a coherent habitat management package – avoidance of night-time working or within two hours of dawn and dusk.

SURVEYS FOR PROTECTED SPECIES: WHAT THE LPA MIGHT HAVE ORDERED

Species: White-clawed Crayfish	
	Worst case scenario: translocation. As part of a coherent set of species plans, translocation may be cited as the only option. This requires clear demonstration that the activity would benefit WCC (Natural England 2022h). This must be within the same catchment to preclude disease spread. Any habitat used in translocation should be capable of supporting a viable set of populations in a habitat (probably new, and developed ahead of removal (Natural England 2022a). Such areas would need to be analogous to the removal areas, and at least as large and safe from damage or loss in the long term. It would need to be managed as part of a long-term, funded plan.
Licensable activities	As a European protected species, licences are required to handle WCC. These are available from the Environment Agency and need securing in advance. If there is need to move WCC under licence, this will be as part of an agreed mitigation plan (Natural England 2022h). Licences can only be given where it is clear that there are justifiable measures for affecting WCC. Any habitat alteration will require a licence. Natural England will only issue a license if the proposal enhances WCC habitats and helps to conserve them.
Monitoring	If mitigation is planned as part of a long-term funded programme, then it normally requires monitoring. Objectives will need to be stated. Timing will meet normal survey methods and frequency and remediation actions are undertaken as a result in line with plan requirements.
Summary	A PEA is required, with field and desk data. Viable field visits are possible May to September, but preferable July to September. A minimum of one visit is required; subsequent monitoring should be spread over two years. The standard survey method is a day-based method. Less efficient methods are also available. Surveys are affected by weather: water flow and levels of run off, turbidity and chemical composition. Wider coverage than the site alone is needed to fully understand potential population impacts. Mitigation needs to assess habitat impacts, and if unavoidable, will require site management. Any translocation requires pre-prepared suitable habitats without pre-existing WCC populations and with long-term security and funded plans in place. Efficacy is to be confirmed. All actions need to be mindful of disease risk and introduction of alien crayfish species.

WCC = White-clawed Crayfish; PEA = Preliminary Ecological Appraisal; ZoI = zone of influence; NE= Natural England; NS = NatureScot; CIEEM = Chartered Institute of Ecological and Environmental Management

Table 5.10 Species guidance: freshwater fish

Species: freshwater fish	
	A range of fish are protected under the Wildlife and Countryside Act 1981: Allis Shad *Alosa alosa*, Twaite Shad *Alosa fallax*, Vendace *Coregonus albula*, and Whitefish *Coregonus lavaretus* – also known as Powan, Gwyniad or Schelly – and Atlantic Sturgeon *Acipenser sturio*. In addition, special areas of conservation (SACs), sites of special scientific interest (SSSIs) or Ramsar sites have features of special interest for freshwater or migratory fish, such as: Atlantic Salmon *Salmo salar*, Bullhead *Cottus gobio*, lamprey (brook, river and sea) (Petromyzontiformes), Spined Loach *Cobitis taenia* and European Eel *Anguilla anguilla*.
Site name and status	Protected fish occur in designated and non-designated habitats across the UK. In England, site status can be gleaned from Natural England's MAGIC website and is a basic part of a PEA desk search.
ZoI	Protected fish occur in a range of waterbodies, both flowing and standing. Because they are susceptible to upstream-derived and off-site pollution, effects of erosion and siltation, habitat loss and movement barriers, in most cases it is reasonable to look at a catchment level in the first instance. This especially applies to those species that are part or wholly migratory.
Identified in pre-application discussions	Pre-application discussions will normally include all aspects of biodiversity, especially protected species. Most LPA discussions will refer to validation checks and these will include fish in a number of areas. If potential protected fish sites or catchments are involved, and they are not in the discussion, this should be queried by LPAs.
The species should be picked up in a PEA	The PEA will include desk survey data sought from Local Record Centres, the National Biodiversity Network and specialist groups as well as field surveys. The walkover element of a PEA will indicate suitable habitat and possible evidence of occurrence. For almost all fish species, absence in brief searches during a PEA visit would not be in any way definitive. Natural England (2022a) cautions against using record centre data that are older than two years. Field level data should be no more than two years old (CIEEM 2019b). Absence from a desk search or lack of initial site evidence is not the same as site absence (Natural England 2022i).
Protected species in and around the site	After using the PEA desk search as a screening for records in and around the site, walked coverage should extend beyond the site boundaries as noted in the ZoI. There is no formal advice for distance in advisory guidance (JNCC 2015).
Age of data attached to the application	Field data are out of date within two years (CIEEM 2019i).

SURVEYS FOR PROTECTED SPECIES: WHAT THE LPA MIGHT HAVE ORDERED | 137

Species: freshwater fish	
	The objective of the survey will determine the methods used and the level of effort, which will also be affected by the time of year and climatic variables (JNCC 2015). Objectives will include: – presence/absence – enough to stimulate a detailed survey, or confirm basic status – assessing abundance and age structure – assessing species population sizes – assessing potential habitat loss and management effects – effects of removing populations in case of partial/total area destruction. Subsequent to any development, a round of monitoring surveys would be expected, with these matching techniques pre-development. These would be as part of a mitigation plan (Natural England 2022i).
Field collection methods	There are individual methods for each species.
Survey timing	This varies with species, but is typically April to October, with short periods best for each species.
Timing of visits in the day	Almost all species are best surveyed in good daylight.
Visit frequency	At least one visit per year, for a minimum of three years.
Survey duration	Depending on the species, at least a 24-hour period.
Weather	Sampling can be affected by flood conditions, water levels, water temperature, light conditions and air temperature.
Light	Light conditions affect visibility and sampling efficacy.
Distance	According to species sampling, it will include the whole river width and sampling lengths of at least 100 m.
How far beyond development boundaries	Fish sampling is within the waterbody, but potential impacts should be evaluated upstream and in the proximate watersheds. There is no formal distance for most species.
Years of data	JNCC and others (Hillman et al. 2003; Pritchard et al. 2021) recommend three years of data; at least one year's worth of visits to establish status.
Limitations	Surveys are affected by: time of year, weather (rain: water levels, sediment, turbidity, water temperature) and surveyor experience.
Suitable licensed and experienced surveyors	Gill netting requires an Environment Agency licence. Salmon and Lamprey electrofishing requires licences in England, Wales, Scotland and Northern Ireland. Sturgeon are fully protected as a European Protected Species, so needs a licence to disturb it or its habitat. Any negative activity would require a European Protected Species mitigation licence.

(continued)

Table 5.10 Species guidance: freshwater fish (*continued*)

Species: freshwater fish	
Determining impacts	Assessing impacts will require knowing the potential extent and form of impact and an indication of the scale of the population using part or the whole area. Impacts will include: – habitat loss and deterioration – fragmentation of habitats or closing dispersal routes between habitat blocks used by fish species – incidental mortality – water regime changes, including oxygen concentrations – water pollution leading to secondary habitat effects – sediment pollution – light pollution – changes in flow regimes – changes in waterbody/river shape and current regimes – changes in bank or in situ vegetation.
Mitigating properly assessed potential impacts	Avoidance of any direct mechanical impact or indirect impact would be the very basic requirements for protected fish. Suitable avoidance actions (Natural England 2022i) would include: – retaining current habitats in situ – changing work times to avoid the spawning period – putting temporary fencing to control sediment or other deleterious runoff entering the system/waterbody – avoidance of night-time working and light use – avoiding working within the waterbody – increasing connectivity to suitable habitat areas with clear evidence of suitability for each species involved. Worst case scenario: translocation. As part of a coherent set of species plans, translocation may be cited as the only option. This requires clear demonstration that the activity would benefit protected species (Natural England 2022i). This must be within the same catchment to preclude disease spread. Any habitat use in translocation should be capable of supporting a viable set of populations in a habitat (probably new), and developed ahead of removal (Natural England 2022i). Such areas would need to be analogous to the removal areas and at least as large and safe from damage or loss in the long term. It would need to be managed as part of a long-term, funded and monitored plan.
Licensable activities	Licences can only be given where it is clear that there are justifiable measures for affecting species. Any habitat alteration will require a licence. Natural England will only issue a license if the proposal enhances fish habitats and helps to conserve them. Most protected fish require Environment Agency licences to catch them. eDNA is currently not a licensable activity.

Species: freshwater fish	
Monitoring	If mitigation is planned as part of a long-term funded programme, then it normally requires monitoring. Objectives will need to be stated. Timing will meet normal survey methods and frequency, and remediation actions undertaken as a result in line with plan requirements.
Summary	A PEA is required, with field and desk data. Viable sampling periods vary with species. Most are July to September. A minimum of one visit is required, with three years of data preferable. Standard survey methods are normally day based. Surveys are affected by weather, water flow and levels of run off, turbidity and chemical composition. eDNA is best used in normal flow periods. Work is still ongoing about the extent to which eDNA confirm density as well as presence of species. Wider coverage than the site alone is needed to fully understand potential population impacts. Mitigation needs to assess habitat impacts, and if unavoidable, will require site management. Any translocation requires pre-prepared suitable habitats without affecting pre-existing populations and with long-term security and funded plans in place. Efficacy of mitigation has yet to be determined. All actions need to be mindful of disease risk and introduction of alien species.

PEA = Preliminary Ecological Appraisal; ZoI = zone of influence; NE = Natural England

Table 5.10a Species guidance: freshwater fish – Whitefish/Vendace

Species: freshwater fish – Whitefish/Vendace	
	Vendace and Whitefish are protected by the Habitats and Species Directive.
Site name and status	Protected fish occur in designated and non-designated habitats across the UK. In England, site status can be gleaned from Natural England's MAGIC website and is a basic part of a PEA desk search.
ZoI	Both species occur in a limited number of lakes. Vendace (*Coregonus Albula*) occurs in two Scottish lakes (Castle Loch and Mill Loch, Lochmaben) and Bassenthwaite and Derwentwater in Cumbria. Because they are susceptible to upstream derived and off-site pollution, effects of erosion and siltation, habitat loss and movement barriers, in most cases it is reasonable to look at a catchment level in the first instance as well as in the lakes.
Identified in pre-application discussions	Pre-application discussions will normally include all aspects of biodiversity, especially protected species. Most LPA discussions will refer to validation checks and these will include fish in a number of areas. If potential protected fish sites or catchments are involved, and they are not in the discussion, this should be queried by LPAs.

(continued)

Table 5.10a Species guidance: freshwater fish – Whitefish/Vendace (*continued*)

Species: freshwater fish – Whitefish/Vendace	
The species should be picked up in a PEA	The PEA will include desk survey data sought from Local Record Centres, the National Biodiversity Network and specialist groups as well as field surveys. The walkover element of a PEA will indicate suitable habitat and possible evidence of occurrence. For almost all fish species, absence in brief searches during a PEA visit would not be in any way definitive. Natural England (2022a) cautions against using record centre data that are older than two years. Field level data should be no more than two years old (CIEEM 2019b). Absence from a desk search or lack of initial site evidence is not the same as site absence (Natural England 2022i).
Protected species in and around the site	After using the PEA desk search as a screening for records in and around the site, walked coverage should extend beyond site boundaries as noted in the ZoI. There is no formal advice for distance in advisory guidance (JNCC 2015).
Age of data attached to the application	Field data are out of date within two years (CIEEM 2019b; Natural England 2022a).
Objectives of surveys	The objective of the survey will determine the methods used and the level of effort, which will also be affected by the time of year and climatic variables (JNCC 2015). Objectives will include: – presence/absence – enough to stimulate a detailed survey or confirm basic status – assessing abundance and age structure – assessing species population sizes – assessing potential habitat loss and management effects – effects of removing populations in case of partial/total area destruction. Subsequent to any development, a round of monitoring surveys would be expected, with these matching techniques pre-development. These would be as part of a mitigation plan (Natural England 2022i).
Field collection methods	Whitefish/Vendace: These are sampled using a combination of gill netting and quantitative hydroacoustics. Each site is pre-surveyed to identify suitable habitat areas and bathymetric profiles in order to correctly set NORDIC gill nets and transect lines for hydroacoustic sampling (JNCC 2015). Hydroacoustic sampling will need to use a calibrated split-beam system. Whitefish/Vendace are sampled under licence between July and September when fish have grown and can be sampled effectively by both methods. Surveys are carried out over 24 hours, starting in daylight, and being repeated under darkness. Transects need to be accurately recorded for subsequent repeat use.

Species: freshwater fish – Whitefish/Vendace	
Survey timing	For Whitefish/Vendace: July to September.
Timing of visits in the day	Whitefish and Vendace sampling is done both in daylight and at night.
Visit frequency	For basic assessments, one visit per year for three years to cover year-to-year variation.
Survey duration	Over a 24-hour period. If time is an issue, night-time work is preferable.
Weather	Avoidance of flood conditions allows comparability between years.
Light	Both species are more active at night.
Distance	Transect and net gills are chosen to fully represent suitable habitats within the waterbodies.
How far beyond development boundaries	Fish sampling is within the waterbody, but potential impacts should be evaluated upstream and in the proximate watersheds.
Years of data	JNCC (2015) recommends three years of data; at least one year's worth of visits to establish status.
Limitations	Surveys are affected by: time of year, weather (rain: water levels, sediment, turbidity, water temperature) and surveyor experience.
Suitable licensed and experienced surveyors	Gill netting requires an Environment Agency licence, and for Whitefish/Vendace, a survey licence. Licensed surveyors are expected to have previous practical experience. Licences to alter habitat are required.
Determining impacts	Assessing impacts will require knowing the potential extent and form of impact and an indication of the scale of the population using part or the whole area. Impacts will include: – habitat loss and deterioration – fragmentation of habitats or closing dispersal routes between habitat blocks used by fish species – incidental mortality – water regime changes, including oxygen concentrations – water pollution leading to secondary habitat effects – sediment pollution – light pollution – changes in flow regimes – changes in waterbody/river shape and current regimes – changes in bank or in situ vegetation.

(continued)

Table 5.10a   Species guidance: Freshwater fish – Whitefish/Vendace (*continued*)

Species: freshwater fish – Whitefish/Vendace	
Mitigating properly assessed potential impacts	Avoidance of any direct mechanical impact or indirect impact would be the very basic requirement for protected fish. Suitable avoidance actions (Natural England 2022i) would include: – retaining current habitats in situ – changing work times to avoid the spawning period – putting temporary fencing to control sediment or other deleterious runoff entering the system/waterbody – avoidance of night-time working and light use – temporary fencing to control WV access to work areas – avoiding working within the waterbody – increasing connectivity to suitable habitat areas, with clear evidence of suitability for each species involved. Worst case scenario: translocation. As part of a coherent set of species plans, translocation may be cited as the only option. This requires clear demonstration that the activity would benefit protected species (Natural England 2022i). This must be within the same catchment to preclude disease spread. Any habitat use in translocation should be capable of supporting a viable set of populations in a habitat (probably new, and developed ahead of removal (Natural England 2022i). Such areas would need to be analogous to the removal areas, and at least as large and safe from damage or loss in the long term. It would need to be managed as part of a long-term, funded and monitored plan.
Licensable activities	Licences can only be given where it is clear that there are justifiable measures for affecting species. Any habitat alteration will require a licence. Natural England will only issue a license if the proposal enhances fish habitats and helps to conserve them. Most protected fish require Environment Agency licences to catch them. eDNA is currently not a licensable activity.
Monitoring	If mitigation is planned as part of a long-term funded programme, then it normally requires monitoring. Objectives will need to be stated. Timing will meet normal survey methods and frequency, and remediation actions undertaken as a result in line with plan requirements.

Species: freshwater fish – Whitefish/Vendace	
Summary	A PEA is required, with field and desk data. Viable sampling periods are July–September. A minimum of 1 visit is required, with three years of data preferable. Standard survey methods are normally day based. Surveys are affected by weather: water flow and levels of run off, turbidity, chemical composition. eDNA is best used in normal flow periods. Work is still ongoing about the extent to which eDNA can be used to confirm density as well as presence of species. Wider coverage than the site alone is needed to fully understand potential population impacts. Mitigation needs to assess habitat impacts, and if unavoidable, will require site management. Any translocation requires pre-prepared suitable habitats without affecting pre-existing populations and with long-term security and funded plans in place. Efficacy of mitigation has yet to be determined. All actions need to be mindful of disease risk and introduction of alien species.

PEA = Preliminary Ecological Appraisal; ZoI = zone of influence; NE = Natural England; WV = Whitefish/ Vendace

Table 5.10b   Species guidance: Allis Shad and Twaite Shad

Species: freshwater fish – Allis Shad and Twaite Shad	
	Allis Shad and Twaite Shad are protected by section 5 of the 1981 Wildlife and Countryside Act.
Site name and status	Protected fish occur in designated and non-designated habitats across the UK. In England, site status can be gleaned from Natural England's MAGIC website and is a basic part of a PEA desk search.
ZoI	Shad occur in flowing waterbodies, and move into estuaries and then rivers to spawn. Because they are susceptible to upstream-derived and off-site pollution, effects of erosion and siltation, habitat loss and movement barriers, in most cases it is reasonable to look at a part catchment level in the first instance. This especially applies as these species are part or wholly migratory.
Identified in pre-application discussions	Pre-application discussions will normally include all aspects of biodiversity, especially protected species. Most LPA discussions will refer to validation checks and these will include fish in a number of areas. If potential protected fish sites or catchments are involved, and they are not in the discussion, this should be queried by LPAs.

(continued)

Table 5.10b Species guidance: Allis Shad and Twaite Shad (*continued*)

Species: freshwater fish – Allis Shad and Twaite Shad	
The species should be picked up in a PEA	The PEA will include desk survey data sought from Local Record Centres, the National Biodiversity Network and specialist groups as well as field surveys. The walkover element of a PEA will indicate suitable habitat and possible evidence of occurrence. For almost all fish species, absence in brief searches during a PEA visit would not be in any way definitive. Natural England (2022a) cautions against using record centre data that are older than two years. Field level data should be no more than two years old (CIEEM 2019). Absence from a desk search or lack of initial site evidence is not the same as site absence (Natural England 2022i).
Protected species in and around the site	After using the PEA desk search as a screening for records in and around the site, walked coverage should extend beyond site boundaries as noted in the ZoI. There is no formal advice for distance in advisory guidance (JNCC 2015).
Age of data attached to the application	Field data are out of date within two years (CIEEM 2019b).
Objectives of surveys	The objective of the survey will determine the methods used and the level of effort, which will also be affected by the time of year and climatic variables (JNCC 2015). Objectives will include: – presence/absence – enough to stimulate a detailed survey, or confirm basic status – assessing abundance and age structure – assessing species population sizes – assessing potential habitat loss and management effects – effects of removing populations in case of partial/total area destruction. Subsequent to any development, a round of monitoring surveys would be expected, with these matching techniques pre-development. These would be as part of a mitigation plan (NE 2022).
Field collection methods	Allis and Twaite Shad: Different methods are used for eggs, juveniles and adult Shad. The reliability of these is in question (JNCC 2015). A) eggs: armed with a micro invertebrate net, sampling of suitable habitat requires starting at the bottom of an area and kicking upstream of the net for 15 seconds. The net is then checked for eggs/debris and recommenced for 25 kicks/30 minutes if no eggs are found. If present, sampling continues 10 m upstream and downstream until no more eggs are found. Sampling only occurs in warmer water and standard flow conditions. Sampling continues for five samples after last eggs are found. Eggs are checked for species via genetic analysis (JNCC 2015).

Species: freshwater fish – Allis Shad and Twaite Shad	
	B) Adults: these are noted at automatic sampling points on known Shad rivers and will be used as indicators of status between points. C) Juveniles: catching occurs in lower reaches of rivers/estuaries, so it is unlikely to occur at most development proposal areas. It involves monthly seine netting across the river on an ebbing neap tide, with three sweeps taken and water height and temperature recorded. It should not be done in poor water or weather conditions (Hillman et al. 2003).
Survey timing	For Shad, sampling times depend on which stage of the lifecycle: – eggs: May and June – adults: April to June – juveniles: netting monthly July to October, with preference for earlier in that period.
Timing of visits in the day	For shad egg sampling, good light and water conditions are critical. Adult and juvenile timing is less critical: adults are counted on counters, and young by netting.
Visit frequency	Single visits for shad, as part of a minimum three year run of data.
Survey duration	Kick sampling for eggs takes around 30 minutes per suitable site. Automatic counting for Shad is part of an ongoing process. Netting juvenile Shad requires three net sweeps once per month on an ebbing neap tide (Hillman et al. 2003).
Weather	Shad runs are affected by water levels and water temperature (Hillman et al. 2003).
Light	Good light is needed for egg sampling by kick sampling.
Distance	Kick sampling for Shad eggs requires around 10 m of suitable habitat.
How far beyond development boundaries	Shad sampling for eggs would be on-site. Automatic counting would refer to the nearest counter, and juvenile counting would most likely be outside of the prospective site.
Years of data	JNCC (2015) and Hillman et al. (2003) recommend three years of data; at least one year's worth of visits to establish status.
Limitations	Surveys are affected by: time of year, weather (rain: water levels, sediment, turbidity, water temperature) and surveyor experience.

(continued)

Table 5.10b Species guidance: Allis Shad and Twaite Shad (*continued*)

Species: freshwater fish – Allis Shad and Twaite Shad	
Suitable licensed and experienced surveyors	Shad netting requires an Environment Agency licence.
Determining impacts	Assessing impacts will require knowing the potential extent and form of impact and an indication of the scale of the population using part or the whole area. Impacts will include: – habitat loss and deterioration – fragmentation of habitats or closing dispersal routes between habitat blocks used by fish species – incidental mortality – water regime changes, including oxygen concentrations – water pollution leading to secondary habitat effects – sediment pollution – light pollution – changes in flow regimes – changes in waterbody/river shape and current regimes – changes in bank or in situ vegetation.
Mitigating properly assessed potential impacts	Avoidance of any direct mechanical impact or indirect impact would be the very basic requirements for protected fish. Suitable avoidance actions (Natural England 2022i) would include: – retaining current habitats in situ – changing work times to avoid the spawning period – putting temporary fencing to control sediment or other deleterious runoff entering the system/waterbody – avoidance of night-time working and light use – avoiding working within the waterbody – increasing connectivity to suitable habitat areas, with clear evidence of suitability for each species involved. Worst case scenario: translocation. As part of a coherent set of species plans, translocation may be cited as the only option. This requires clear demonstration that the activity would benefit protected species (Natural England 2022i). This must be within the same catchment to preclude disease spread. Any habitat use in translocation should be capable of supporting a viable set of populations in a habitat (probably new, and developed ahead of removal (Natural England 2022i)). Such areas would need to be analogous to the removal areas and at least as large and safe from damage or loss in the long term. It would need to be managed as part of a long-term, funded and monitored plan.

SURVEYS FOR PROTECTED SPECIES: WHAT THE LPA MIGHT HAVE ORDERED | 147

Species: freshwater fish – Allis Shad and Twaite Shad	
Licensable activities	Licences can only be given where it is clear that there are justifiable measures for affecting species. Any habitat alteration will require a licence. Natural England will only issue a license if the proposal enhances fish habitats and helps to conserve them. Most protected fish require Environment Agency licences to catch them. eDNA is currently not a licensable activity.
Monitoring	If mitigation is planned as part of a long-term funded programme, then it normally requires monitoring. Objectives will need to be stated. Timing will meet normal survey methods and frequency and remediation actions undertaken as a result in line with plan requirements.
Summary	A PEA is required, with field and desk data. Viable sampling periods vary with species. Eggs are sampled May to June, adults April to June and juveniles July to October. A minimum of one visit is required, with three years of data preferable. Standard survey methods are normally day based. Surveys are affected by weather, water flow and levels of run off, turbidity and chemical composition. eDNA is best used in normal flow periods. Work is still ongoing about the extent to which eDNA confirm density as well as presence of species. Wider coverage than the site alone is needed to fully understand potential population impacts. Mitigation needs to assess habitat impacts, and if unavoidable, will require site management. Any translocation requires pre-prepared suitable habitats without affecting pre-existing populations and with long-term security and funded plans in place. Efficacy of mitigation has yet to be determined. All actions need to be mindful of disease risk and introduction of alien species.

PEA = Preliminary Ecological Appraisal; ZoI = zone of influence; NE= Natural England

Table 5.10c   Species guidance: Sturgeon

Species: freshwater fish – Sturgeon	
	Sturgeon are protected the 1981 Wildlife and Countryside Act and Atlantic sturgeon are European protected species (EPS) protected under the Conservation of Habitats and Species Regulations 2017.
Site name and status	Protected fish occur in designated and non-designated habitats across the UK. In England, site status can be gleaned from Natural England's MAGIC website and is a basic part of a PEA desk search.

(continued)

148 | PROTECTED SPECIES AND BIODIVERSITY

Table 5.10c Species guidance: Sturgeon (*continued*)

Species: freshwater fish – Sturgeon	
ZoI	Sturgeon formerly occurred in a range of water flowing waterbodies, and rank as a vagrant in the UK. Because they are susceptible to upstream-derived and off-site pollution, effects of erosion and siltation, habitat loss and movement barriers, in most cases it is reasonable to look at a catchment level in the first instance.
Identified in pre-application discussions	Pre-application discussions will normally include all aspects of biodiversity, especially protected species. Most LPA discussions will refer to validation checks and these will include fish in a number of areas. If potential protected fish sites or catchments are involved and they are not in the discussion, this should be queried by LPAs.
The species should be picked up in a PEA	The PEA will include desk survey data sought from Local Record Centres, the National Biodiversity Network and specialist groups as well as field surveys. The walkover element of a PEA will indicate suitable habitat and possible evidence of occurrence. For almost all fish species, absence in brief searches during a PEA visit would not be in any way definitive. Field level data should be no more than two years old (CIEEM 2019b). Absence from a desk search or lack of initial site evidence is not the same as site absence (Natural England 2022i).
Protected species in and around the site	After using the PEA desk search as a screening for records in and around the site, walked coverage should extend beyond site boundaries as noted in the ZoI. There is no formal advice for distance in advisory guidance (JNCC 2015).
Age of data attached to the application	Field data are out of date within two years. (CIEEM 2019b; Natural England 2022a).
Objectives of surveys	The objective of the survey will determine the methods used and the level of effort, which will also be affected by the time of year and climatic variables (JNCC 2015). Objectives will include: – presence/absence – enough to stimulate a detailed survey, or confirm basic status – assessing abundance and age structure – assessing species populations sizes – assessing potential habitat loss and management effects – effects of removing populations in case of partial/total area destruction. Subsequent to any development, a round of monitoring surveys would be expected, with these matching techniques pre-development. These would be as part of a mitigation plan (NE 2022j).

Species: freshwater fish – Sturgeon	
Field collection methods	In the UK, Sturgeon are incidental take species. There is no formal sampling programme, but it may be detected using bioacoustics methods (Flowers 2013) and also via eDNA (Tang 2020). Care needs to be taken over risk of disease transfer between catchment areas.
Survey timing	Sturgeon breed in spring to summer (EC LIFE) in Europe but no longer in the UK.
Timing of visits in the day	Sturgeon sampling is not well established.
Visit frequency	Sturgeon, if suspected, would require targeted use of sonar sampling.
Survey duration	Sturgeon sampling needs linking to potential numbers (EC LIFE).
Weather	Sturgeon sampling is affected by water levels and water temperature.
Light	Light is not a key factor for sturgeon.
Distance	The minimum distance for very basic sampling is 100 m.
How far beyond development boundaries	There is no formal distance for sturgeon.
Years of data	At least one year of data.
Limitations	Surveys are affected by: time of year, weather (rain: water levels, sediment, turbidity, water temperature) and surveyor experience.
Suitable licensed and experienced surveyors	Sturgeon are fully protected as a European Protected Species, so require a licence to disturb them or their habitat. Any negative activity would require a European Protected Species mitigation licence.
Determining impacts	Assessing impacts will require knowing the potential extent and form of impact and an indication of the scale of the population using part of or the whole area. Impacts will include: – habitat loss and deterioration – fragmentation of habitats or closing of dispersal routes between habitat blocks used by fish species – incidental mortality – water regime changes, including oxygen concentrations – water pollution leading to secondary habitat effects – sediment pollution – light pollution – changes in flow regimes – changes in waterbody/river shape and current regimes – changes in bank or in situ vegetation.

(continued)

Table 5.10c Species guidance: Sturgeon (*continued*)

Species: freshwater fish – Sturgeon	
Mitigating properly assessed potential impacts	Avoidance of any direct mechanical impact or indirect impact would be the very basic requirements for protected fish. Suitable avoidance actions (Natural England 2022i) would include: – retaining current habitats in situ – changing work times to avoid the spawning period – putting temporary fencing up to control sediment or other deleterious runoff entering the system/waterbody – avoidance of night-time working and light use – avoiding working within the waterbody – increasing connectivity to suitable habitat areas, with clear evidence of suitability for each species involved.
Licensable activities	As a European Protected species, Natural England licences are required to handle sturgeon. These need securing in advance. If there is need to move fish under licence, this will be as part of an agreed mitigation plan (Natural England 2022i). eDNA is currently not a licensable activity.
Monitoring	If mitigation is planned as part of a long-term funded programme, then it normally requires monitoring. Objectives will need to be stated. Timing will meet normal survey methods and frequency, and remediation actions undertaken as a result should be in line with plan requirements.
Summary	A PEA is required, with field and desk data. There is no simple sampling period. A minimum of one visit is required, with three years of data preferable. eDNA is best used in normal flow periods. Work is still ongoing about the extent to which eDNA confirm density as well as presence of species. Wider coverage than the site alone is needed to fully understand potential population impacts. Mitigation needs to assess habitat impacts, and if unavoidable, will require site management. Any translocation requires pre-prepared suitable habitats without affecting pre-existing populations and with long-term security and funded plans in place. Efficacy of mitigation has yet to be determined. All actions need to be mindful of disease risk and introduction of alien species.

PEA = Preliminary Ecological Appraisal; ZoI = zone of influence; NE = Natural England

Table 5.11 Species guidance: ancient woodland, ancient trees and veteran trees

Species: ancient woodland, ancient trees and veteran trees	
Site name and status	Ancient woodland, ancient trees and veteran trees (AVT) occur in designated and non-designated habitats across the UK. In England, site status can be gleaned from Natural England's MAGIC website and is a basic part of a PEA desk search.
ZoI	AVT occur in a range of locations, and their status may well be as part of a wider woodland, parkland or wood pasture, so AVT need to be seen in context (NE 2022). NE (2022) requires a local planning authority to use an assessment guide included in the NE guidance 'Assessment guide: ancient woodland, ancient and veteran trees', which documents the ZoI.
Identified in pre-application discussions	Pre-application discussions will normally include all aspects of biodiversity, especially protected species and AVT. Most LPA discussions will refer to validation checks and these will include AVT in a number of areas. If potential AVT sites are involved, and they are not in the discussion, this should be queried by LPAs.
The species should be picked up in the PEA	An AVT PEA will include desk survey data sought from Local Record Centres, the National Biodiversity Network, Natural England's Ancient Woodland Inventory (NWI) (https://naturalengland-DEFRA.opendata.arcgis.com/datasets/DEFRA::ancient-woodland-england/about), Woodland Trust's Ancient Tree inventory (ATI) (https://ati.woodlandtrust.org.uk/) and Natural England's wood pasture and parkland inventory and specialist groups as well as field surveys. The walkover element of a PEA will indicate suitable habitat. Absence in brief desk searches would not be definitive, as AVTs smaller than 2 ha are often omitted. These areas may be found via Local Record Centres (Natural England 2022j). Full species lists should be sought from Local Record Centres and the National Biodiversity Network, while being mindful of the limits of both data sources. There is no caution against old (more than two years) data, but NE does note that the NWI and ATI are always being updated. Absence from a desk search or lack of initial site evidence is not the same as AVT absence (Natural England 2022j).
Protected species in and around the site	After using the PEA desk search as a screening for records in and around the site, walked coverage should extend beyond site boundaries as noted in the ZoI.

(continued)

Table 5.11 Species guidance: ancient woodland, ancient trees and veteran trees (*continued*)

Species: ancient woodland, ancient trees and veteran trees	
Age of data attached to the application	Data for AVT are being updated. For trees, the two-year record limit will not apply. For all other taxa, the two-year limit (CIEEM 2019b) applies.
Objectives of surveys	Natural England (2022j) states that there should be enough suitable evidence to make a decision. This will be based on map and field survey. Objectives will include: – establishing for ancient woodland that there has been woodland present continuously since 1600; that includes open areas which are part of the dynamic and fabric of woodland (Natural England 2022j). – identifying potential veteran or ancient trees in a survey area – establishing the status of ancient and veteran trees and their linked biodiversity interest – establishing the ecological linkages/contexts of AVT – establishing the biodiversity and species communities associated with AVT, especially the important or protected species associated with them and their species population sizes in case of potential impacts – assessing potential habitat loss and management effects in a development in or near AVT – for non-tree species, the presence of species of conservation importance within AVT areas and their link to components of the AVT. Subsequent to any development, a round of monitoring surveys would be expected, with these matching techniques pre-development. These would be as part of a mitigation plan (NE 2022). Note that all AVT are categorised as irreplaceable (Natural England 2022j).
Field collection methods	Natural England (2022j) requires tree surveys (BS 5837 (2012): https://knowledge.bsigroup.com/products/trees-in-relation-to-design-demolition-and-construction-recommendations/standard) and ecological surveys. The latter states that surveyors should use Natural England (2022j) for survey guidance. Note that in view of the problems inherent in the January 2022 SA guidance, alternatives, such as that presented in this guide, should be used instead for protected species.

Species: ancient woodland, ancient trees and veteran trees	Note also that BS 5837 is focused on development and tree condition. Veteran trees have a few mentions, and ancient woodland has none in BS 5837. Biodiversity has two generic entries. This means that it is important that effective field surveys for a range of taxa are included, in addition to BS 5837 reports. In addition to BS 5837, surveyors should use Fay and de Berker (1997) and Fay (2007), which go through the categorisation of veteran and ancient trees in detail and describe the attributes that will help classify trees that have potentially not been put under the banner of AVT before. Confusingly, the Forestry Commission/ Natural England (FC/ NE) assessment guide that LPAs are required to follow according to the Natural England (ED 2022j) AVT guidance lists a range of additional generic surveys, such as soil surveys, hydrological surveys and historic environment surveys, without any reference or web links. Some of these can be checked in Hill et al. (2005).

1. BS 5837 (2012).
   The objective of a BS 5837 survey is to produce a map and report, showing all trees and their form (diameter at breast height (DBH), crown and height/condition) as well as ground flora on a topographical map. It should be accompanied by a soil assessment, including soil structure, composition and pH. The survey will include individual ratings of mapped trees according to a standard categorisation which leans towards risk and tree removal after an evaluation of potential longevity. The potential root protection area (RPA) is given for each tree based on a radius of twelve times the stem diameter. This indicates the minimum area to preclude root damage. Note that all AVT components are assumed to be irreplaceable habitat (NE 2022), so that tree longevity is not a critical component of a BS 5837-based survey. RPA is needed where physical disturbance may be possible as part of a development.
2. Fay (2007) and Fay and de Berker (1997). The authors work through the size, feature and shape attributes that qualify individual trees for veteran and ancient status (VAS). These vary between species, so that it is important that individual trees are evaluated. Note that the use of Fay (2007) and Fay and de Berker (1997) would be in addition to BS 5837 which does not go into anything like as much detail on VAS categorisation. They do not cover other species in any systematic way, so a VAS survey would still need to be supplemented by surveys for other taxa.
3. Other survey groups. Details for other protected species groups expected in AVT are covered in other tables in this chapter and should be consulted for methods, times and limitations.

(continued)

Table 5.11 Species guidance: ancient woodland, ancient trees and veteran trees (*continued*)

Species: ancient woodland, ancient trees and veteran trees	
Survey timing	1. BS 5837 (2012): The structural assessment element, soil, RPA and general context can be undertaken at any stage of the year. Floral and species surveys are much more constrained. If using the basic phase 1 method (JNCC 2010), this is limited to April to October in most areas, with floral composition varying across the season so the limits of single visits are understood. 2. For Veteran and Ancient Trees (VAS) assessment, all times of the year are possible. 3. Other species or groups will have their own optimal visit times, and working outside of any of these may seriously compromise the reliability of the data.
Timing of visits	1. BS 5837 (2012) can be undertaken in daylight hours. 2. VAS surveys require standard daylight conditions. 3. Other surveys for species that may be expected will have own time requirements.
Visit frequency	1. BS 5837 (2012): for effective status assessments, a minimum of one visit will be enough, but to fully understand the associated biodiversity, several visits in suitable periods would be needed. 2. VAS surveys can be undertaken over the course of one series of visits to establish VAS status. 3. Other groups will have specific timing and frequency requirements.
Survey duration	1. BS 5837 (2012): there is no set formula, and time needed would vary with site type and complexity. 2. There is no set time for VAS assessment. 3. There may be different requirements for individual species surveys.
Weather	1. BS 5837 (2012): mapping for a tree survey is not limited by weather, other than in Health and Safety Executive terms. Gathering data for other taxa is weather dependent. 2. For VAS purposes, there are no obvious limits, other than in Health and Safety Executive terms. 3. For many species, weather is a key limiting factor in detection and is critical to record in interpreting species surveys and potential impacts.
Light	1. BS 5837 (2012) surveys can be undertaken in daylight hours. 2. VAS surveys requires daylight hours. 3. Light conditions will influence many of the species found in AVT.

SURVEYS FOR PROTECTED SPECIES: WHAT THE LPA MIGHT HAVE ORDERED | 155

Species: ancient woodland, ancient trees and veteran trees	
Distance	1. BS 5837 (2012). The use of RPA is a proxy buffer. Natural England (2022j) takes the RPA up from 12 to 15 times the tree stem diameter. In addition, it recommends a 15 m minimum RPA and a buffer of 5 m from the edge of a tree's canopy if that would be greater than the 15 m RPA. 2. VAS survey describes the tree and also estimates the RPA based on BS 5837 (2012). 3. Survey distance varies between species and will be important in interpreting potential impacts and buffer distances.
How far beyond development boundaries	1. BS 5837 (2012). There is no formal distance discussion, other than RPA. 2. For VAS survey, there is no formal distance discussion, other than RPA. 3. This varies with individual species and groups.
Years of data	1. BS 5837 (2012). For fixed entities like trees, the basic assessment can be made within a year. For some species, more than one year's data may be needed to appraise species use, especially where there are water level variations between seasons. 2. One year's data are normal for VAS survey. 3. For most species, this will be two years (Natural England 2022a, CIEEM 2019b).
Limitations	1. BS 5837 (2012) surveys are affected by time of year and surveyor experience. 2. VAS surveys are affected by surveyor experience. 3. There are many limiting factors in individual species or groups. These need evaluation before assessing potential impacts. If incorrectly determined, or not validated, they may preclude meaningful impact assessment.
Suitable licensed and experienced surveyors	1. It is possible to gain accreditation for BS 5837 (2012). There is no formal licence requirement. 2. VAS survey would be expected to be carried out by qualified arboriculturists. 3. Many species expected in and around AVT may well require licenses.
Determining impacts	Assessing impacts will require knowing the potential extent and form of impact and an indication of the scale of the population using part of or the whole area.

(continued)

Table 5.11 Species guidance: ancient woodland, ancient trees and veteran trees (*continued*)

Species: ancient woodland, ancient trees and veteran trees	
	Direct Impacts on AVT will include: – damaging or destroying all or part of them (including their soils, ground flora or fungi) – damaging roots and understorey (all the vegetation under the taller trees) – damaging or compacting soil – damaging functional habitat connections, such as open habitats between the trees in wood pasture and parkland – increasing levels of air and light pollution, noise and vibration – changing the water table or drainage – damaging archaeological features or heritage assets – changing the woodland ecosystem by removing the woodland edge or thinning trees, causing greater wind damage and soil loss – indirect effects of development can also cause the loss or deterioration of ancient woodland, ancient and veteran trees by: – breaking up or destroying working connections between woodlands, or ancient trees or veteran trees, affecting protected species, such as bats or wood-decay insects – reducing the amount of semi-natural habitats next to ancient woodland that provide important dispersal and feeding habitat for woodland species – reducing the resilience of the woodland or trees and making them more vulnerable to change – increasing the amount of dust, light, water, air and soil pollution – increasing disturbance to wildlife, such as noise from additional people and traffic – increasing damage to habitat, for example trampling of plants and erosion of soil by people accessing the woodland or tree root protection areas – increasing damaging activities like fly-tipping and the impact of domestic pets – increasing the risk of damage to people and property by falling branches or trees requiring tree management that could cause habitat deterioration – changing the landscape character of the area.

Species: ancient woodland, ancient trees and veteran trees	
Mitigating properly assessed potential impacts	Avoidance of any mechanical impact or indirect impact would be the very basic requirement for AVT retention. Note that Natural England (2022j) views granting planning permission as possible only when wholly exceptional reasons are demonstrated. The most basic element would be to avoid the AVT area, including the buffer area. Natural England (2022j) mitigation options include: – putting up screening barriers to protect ancient woodland or ancient and veteran trees from dust and pollution – measures to reduce noise or light – designing open space to protect AVTs – rerouting footpaths and managing vegetation to deflect trampling pressure away from sensitive locations – creating buffer zones of semi-natural habitats. Compensation measures would include: – create new native woodland or wood pasture and allow for natural regeneration – improve the condition of the woodland – remove invasive species – restore or manage other ancient woodland, including plantations on ancient woodland sites, wood pasture and parkland – connect woodland and AVTs separated by development with green bridges, tunnels or hedgerows – produce long-term management plans for new woodland and ancient woodland, including deer management – manage AVTs to improve their condition plant or protect individual trees that could become AVTs in future. The compensation strategy should include monitoring the ecology of the site over an agreed period. Natural England (2022j) notes that it is possible to plant additional woodland, but it is NOT a direct replacement for lost/damaged AVTs. It is also not possible to relocate AVTs in part or whole. Natural England (2022j) indicates that restoration/improvement of AVTs can partially compensate for loss or damage to AVT by: – improving and restoring plantations on ancient woodland sites – improving the management of nearby ancient woodland sites and connecting them better to semi-natural habitat – improving the condition of important features of ancient woodland – improving access for management purposes.

(continued)

Table 5.11 Species guidance: ancient woodland, ancient trees and veteran trees (*continued*)

Species: ancient woodland, ancient trees and veteran trees		
		In addition, proposals can partially compensate for loss or damage to wood pasture by restoring semi-natural open habitat, managed by grazing, with open grown trees. Proposals can partially compensate for the loss or deterioration of AVTs by planting: – young trees of the same species with space around each one to develop an open crown – new trees near to the trees they are replacing. Individual species within AVT may well need their own requirements, and entries for these should be checked as well.
	Licensable activities	Alteration of AVT is not an activity attracting licence requirements. Some of the survey types to assess AVT ecological interests may well need licenses, and should be checked.
	Monitoring	If AVT mitigation is planned as part of a long-term funded programme, then it normally requires monitoring. Objectives will need to be stated. Timing will meet normal survey methods and frequency, and remediation actions undertaken as a result in line with plan requirements.
	Summary	A PEA is required, with field and desk data. A minimum of one visit is required, and subsequent monitoring should be spread over two years. The standard survey method is a day-based method. AVT BS 5837 and VAS surveys are not affected by time of year. Species surveys are more exacting and have detailed requirements for time, time of year and weather. Wider coverage than the site alone is needed to fully understand potential AVT impacts. If changes are deemed acceptable, mitigation needs to assess habitat impacts, and if unavoidable, it will require site management. Any actions need long-term security and funded plans in place.

AVT = Ancient woodland, ancient trees and veteran trees; RPA = root protection area; VAS = veteran and ancient status; PEA = Preliminary Ecological Appraisal; ZoI = zone of influence; NE= Natural England

Table 5.12 Species guidance: protected plants, fungi and lichens

Species: protected plants, fungi and lichens	
Site name and status	PPFL occur in designated and non-designated habitats across the UK. In England, site status can be gleaned from Natural England's MAGIC website and is a basic part of a PEA desk search.
ZoI	PPFL occur in a range of mainly terrestrial environments, but some, such as aquatic macrophytes, occur in very different environments, so the ZoI will vary strongly between PPFL types. PPFL do not exist in isolation, but are influenced by the conditions in the habitats around them (Hill et al. 2005). This is especially so for fungi (Sheldrake 2020). As in all desk searches, a 2 km search zone should be used.
Identified in pre-application discussions	Pre-application discussions will normally include all aspects of biodiversity, especially protected species. Most LPA discussions will refer to validation checks and these will include protected PPFL in a number of areas. If potential PPFL sites are involved, and they are not in the discussion, this should be queried by LPAs.
The species should be picked up in the PEA	The PEA will include desk survey data sought from Local Record Centres, the National Biodiversity Network (NBN) and specialist groups as well as field surveys. The walkover element of a PEA will indicate suitable habitat and possible evidence of occurrence of PPFL. Absence in brief searches during a PEA visit would not be definitive, due to the levels of effort, timing issues and weather issues related to detailed searches. Because of the rarity and protected status of some PPFL, standard desk searches may well draw blanks, so that specialist groups or the specialists in the statutory agencies may need approaching where it is thought some PPFL might occur or when updates for recent status are sought. A PEA desk search should pick up UK BAP priority habitats and Annex 1 habitats, and their associated species interests (UK Habitat 2020), so that there will be indications of potential interests on sites and their habitat types, in addition to the JNCC (2010) phase 1 categories. Natural England (2022) cautions against using Local Record Centre data that are older than two years. Field level data should be no more than two years old (CIEEM 2019). Absence from a desk search or lack of initial site evidence is not the same as site absence (NE 2022). For PPFL, some data will be older than two years, and need to be used with caution.

(continued)

Table 5.12 Species guidance: protected plants, fungi and lichens (*continued*)

Species: protected plants, fungi and lichens	
	Note that PPFL are covered under a range of designations: – European Protected Species covered by schedule 5 of the Conservation of Habitats and Species Regulations 2017: 9 species of plant – Wildlife and Countryside Act 1981 schedule 8: 186 species, including 4 fungi, 31 lichens, 106 higher plants and 45 lower plants – NERC Act 2006: 250 species of lower plants and 152 species of higher plants.
Protected species in and around the site	After using the PEA desk search as a screening for records in and around the site, walked coverage should extend beyond site boundaries as noted in the ZoI.
Age of data attached to the application	Field data are out of date within two years (CIEEM 2019b). This applies to individual species, but not to the habitat types.
Objectives of surveys	The objective of the survey will determine the methods used and the level of effort, which will also be affected by the time of year and climatic variables. Because some PPFL are also European Protected Species, any alteration/destruction will be illegal. As a result, objectives will include detailed surveys, while others may do an initial survey prior to more detailed work. Objectives will include: – presence/absence – enough to stimulate a detailed survey, or confirm basic status – assessing which areas on a site are used by PPFL – assessing species population sizes – assessing potential habitat loss and management effects – assessing potential effects of removing populations in case of partial/total area destruction. Subsequent to any development, a round of monitoring surveys would be expected, with these matching techniques pre-development. These would be as part of a mitigation plan (Natural England 2022j).
Field collection methods	There are a range of methods available for PPFL. General PPFL surveys will be based on assessing known locations and will effectively be a form of monitoring. For that reason, it is important that pre-existing methods are stated, along with their limitations and assumptions (BS 42020 (BSI 2013)). Because of the relative rarity of protected species, specialists will be needed to carry out the surveys, as some species, primarily fungi and lichens, need detailed differentiation and character recognition. Almost all of these will have their own specialised survey methods or group survey methods tailored from standard vegetation sampling.

Species: protected plants, fungi and lichens	
	Many species will tend to have their own best time for survey, and where one or more species occur on a site, these may not all be visible at the same time, so survey times and methods may vary across the seasons. 1. **Protected fungi** Four species are protected under schedule 8 of the 1981 Wildlife and Countryside Act: Sandy Stilt Puffball *Battarrea phalloides*, Royal Bolete *Butyriboletus regius* (formerly *Boletus regius*), Bearded Tooth *Hericium erinaceus* and Oak Polypore *Piptoporus quercinus*. The Sandy Stilt Puffball occurs periodically in a very limited number of sandy areas and has no formal field method, other than to search areas in late summer/early autumn in areas where it has previously occurred. New sites are often found as established sites are barren for a few years (Suffolk 2021). The Royal Bolete has a few records in Hampshire only (https://www.first-nature.com/fungi/butyriboletus-regius.php) under trees in calcareous areas. The bearded tooth occurs on old beech trees, and also oak and birch, with one record per year in the last decade (NBN 2022). There is no formal search method for the fruiting body which is found July to November. Oak Polypore is limited to veteran oak trees and has fewer than four records per year (NBN 2022). As with the others, there is no formal method, and its presence at sites varies between years. Checking possible candidate veteran trees July to August is the preferred option (https://naturebftb.co.uk/wp-content/uploads/2022/01/Oak-polypore-species-account.FINAL_.pdf). Hill et al. (2005) set out a range of quadrat, transect and fixed-point photograph recording for fungi. The four protected species are not ready candidates for any of these, other than providing photograph records. 2. **Protected lichens** A total of 32 species of lichens are protected under schedule 8 of the Wildlife and Countryside Act 1981. As these occur in a range of different environments, there is no proprietary method for each. The scanning coverage by the PEA should have indicated a possible selection of species in or near the site, and the lichens may well be at a single location site, such as a tree, an open exposure or mountain top. As lichens can be difficult to identify beyond the most obvious species (Gilbert 2000; Dobson 1992; Hill et al. 2005), surveys will often require specialist inputs.

(continued)

Table 5.12 Species guidance: protected plants, fungi and lichens (*continued*)

Species: protected plants, fungi and lichens	
	In order to assess the value of an initial survey, the following basics need to be undertaken: A) habitat description: as lichens can often grow in particular conditions or microhabitats, these need describing, as any changes to these linked to a development will risk their longer-term status. This also provides the basis for monitoring and indicators for change that would need responding to. B) colony size and extent: depending on the substrate used, colonies can be measured or mapped to establish as a baseline. Where a ground-based colony is beyond the standard 1 m² quadrat size, then a suitably controlled (to preclude distortions) fixed-point photograph with a suitable scalar will help for future monitoring. C) Quadrats: where colonies do not grow on vertical habitats, and the site is suitable for use of a repeat quadrat without risk of damage to the lichens, a 1 m² quadrat, subdivided into 25 cm sub-squares would be adequate to record the coverage and allow repeat measurements. As lichens are very sensitive to chemical changes, care should be taken to use either steel or plastic markers for subsequent relocation. Use of a buried transponder will help relocate a ground-based quadrat. It is possible to use suitable sized smaller quadrats on vertical surfaces. In all cases, basic details of species, condition and fruits should be noted and are best supported by photographs. 3. Protected plants With 151 non-fungi and lichens covered by the 1981 Wildlife and Countryside Act, and 402 lower and higher plants covered by the 2006 NERC Act, both aquatic and non-aquatic, there is no simple one size fits all set of protocols. In addition, there is no ready basis for Natural England (2022k) to ask LPAs to check if they think that any of these 400 plus species may be on the development site, or possibly affected by the proposal. That is clearly a matter for the developer and their ecologists.

Species: protected plants, fungi and lichens	
	With such a wide range of species and habitats covered by the Acts, and species that flower or grow across the year, there is no one best method. The most suitable method to use depends on the objectives for the survey, and the species. The phase 1 PEA survey should provide a basic habitat map and, along with the limited desk search, indicate the most likely protected species to be found and the areas in which these might occur. This will then narrow down the range of possible methods and times/conditions in which to work, and possible limitations. For example, where a widespread, but protected, species such as Bluebell *Hyacinthoides non-scripta* occurs on part of the site, then mapping the extent of this in peak flowering time and noting its presence within adjacent woodland/hedgerow/heathland (<1,000 m) would provide a basis for assessing potential impacts and mitigation options. By contrast, European Protected Species such as Early Gentian *Gentianella anglica* tend to be found in localised stands on a limited range of sites. If picked up by the PEA, initial walkover checks in the early spring would demarcate the main blocks or stands of the plant and indicate the key areas plus a buffer zone. Detailed surveys would indicate the number of plants and also indicate the condition of the supporting plant communities. Where records are more ambivalent, and a PEA walkover is done outside of the main plant growth season, either JNCC phase 1 or UK Habitat (2020) categories will describe the habitats and likely species of plants and areas to focus on. Note that, with so many species having their own specific growth or flowering periods, a single visit to a previously unrecorded site is likely to underestimate the floral composition of the site and may miss some of the protected species on the site. This should be noted in the limitations section. When looking for protected species in more detail, several options are available. – Look-see method: this is a simple walk through of a site, visiting each of the apparent habitats/communities and recording the number of target species in each area. This normally includes a timed element to begin to standardise the coverage and the route and species are mapped, allowing a later repeat comparison. Issues with observer bias, identification issues and timing make inter-year comparisons or trend statements problematic. This also means later post-development monitoring data are hard to interpret.

(continued)

Table 5.12 Species guidance: protected plants, fungi and lichens (*continued*)

Species: protected plants, fungi and lichens	
	– Systematic counts: this is a more targeted counting of a more spatially limited population(s) and will involve mapping and use of grids and potentially buried permanent markers for subsequent monitoring at suitable time periods within the year. It is normally accompanied by photorecording.
Survey timing	1. Fungi. For the four protected species, the best time is individual (see above) but generally late summer to early autumn. 2. Lichens: there is no formal time. As lichens can be cryptic, in woodlands after leaf fall provides better light, and generally they are more visible in wet in late autumn to early spring when secondary vegetation is absent. 3. Protected plants: each species will have its own best time for visits.
Timing of visits in the day	1. Fungi: most species are visible earlier in the day when there is least chance of damage. 2. Lichens: for cryptic species, good light is critical, so avoidance of early and late in the day is best. 3. Protected plants: few are ephemeral, so the main part of the day is suitable.
Visit frequency	1. Fungi: given that the four protected species are periodic and ephemeral, a single visit in the 'right' period may not be enough to establish presence or absence. 2. Lichens: unless the species is found in an ephemeral site, such as a decaying tree host, one thorough visit will be enough, but for monitoring purposes a formal check every three years is needed (Hill et al. 2005), with a brief annual check being ideal. 3. Protected plants: at least one visit is needed, depending on the species.
Survey duration	1. Fungi: survey duration depends on the site and weather conditions. 2. Lichens: depending on the status of existing knowledge, survey times will vary between sites. 3. Protected plants: this will depend on the patchiness, scale and purpose of the survey.
Weather	1. Fungi: the four protected species tend to favour cooler, wetter periods, so their potential occurrence is influenced by weather. 2. Lichens: they tend to be more visible in wetter conditions. 3. Protected plants: many are affected by weather conditions, especially dry growing seasons which may affect fruiting/flowering or emergence.

Species: protected plants, fungi and lichens	
Light	1. Fungi: good light conditions are required to discover fungi.
2. Lichens: given the complexities recognising some species, good light conditions are essential.
3. Protected plants: most species are visible, but some, such as mosses, may require good conditions in situ for species determination.
Distance	1. Fungi: for many species, sites will be individual locations. For tree users, a 100 m radius would be prudent.
2. Lichens: Like fungi, many lichens are patchily distributed, and a minimum 100 m radius would be prudent, although there is nothing mandated (Hill et al. 2005).
3. Protected plants: as there is a very wide range of species in this category, a minimum 100 m radius would be prudent.
How far beyond development boundaries	1. Fungi: for many species, sites will be individual locations. For tree users, a 100 m radius would be prudent.
2. Lichens: Like fungi, many lichens are patchily distributed, and a minimum 100 m radius would be prudent, although there is nothing mandated (Hill et al. 2005).
3. Protected plants: a minimum 100 m radius would be prudent, although there is nothing mandated (Hill et al. 2005).
Years of data	1. Fungi: at least one year's worth of visits to establish status in principle, mindful of the patchiness of temporal occurrence.
2. Lichens: at least one year's worth of visits to establish status.
3. Protected plants: at least one year's worth of data to establish status. Note the limitations imposed by weather, and potentially grazing where flowers/seeds can be removed, hampering finding/identification.
Limitations	1. Fungi: Surveys are affected by: time of year, weather and surveyor experience.
2. Lichens: surveys are affected by time of year, light, weather and surveyor experience.
3. Protected plants: many species are visible for only part of the year, so that timing is critical, whilst some species of grass are only identifiable when flowering or fruiting (Hill et al. 2005). Species such as Early Gentian may fluctuate strongly between years, and others may have prolonged periods of dormancy. As with lichens and fungi, experienced surveyors are critical in finding and identifying protected plants. As size decreases, the likelihood of species determination and repeatability reduces (Hill et al. 2005). Grazing pressure may be a serious limitation in some locations.

(continued)

Table 5.12 Species guidance: protected plants, fungi and lichens (*continued*)

Species: protected plants, fungi and lichens	
Suitable licensed and experienced surveyors	1–3. Many species of protected fungi, lichen and plants will require a license for surveys, which may involve disturbance (Natural England 2022k).
Determining impacts	Assessing impacts will require knowing the potential extent and form of impact and an indication of the scale of the species using part of or the whole area. Impacts will include: – habitat loss and deterioration – fragmentation of habitats or closing dispersal routes between habitat blocks – incidental mortality – management regime changes – soil water status changes – water flow or status changes – direct or secondary pollution. NB: as time will have elapsed from initial surveys, pre-construction surveys will be recommended for all areas thought to contain or be likely to contain protected species and take place <3 months ahead of planned commencement.
Mitigating properly assessed potential impacts	Avoidance of any mechanical impact or indirect impact would be the very basic requirements for PPFL retention. Suitable avoidance actions (NS 2020a; NE 2022k) would include: – retaining current habitats and protected species in situ – changing methods of working to avoid the active season – temporary fencing to control access to work areas – avoiding introduction of alien plants – increasing connectivity to suitable habitat areas with clear evidence of suitability for each species involved – managing areas within the season to make them suitable as part of a coherent habitat management package. Worst case scenario: translocation. As part of a coherent set of species plans, translocation may be cited as the only option. This requires clear demonstration that the activity would benefit the particular species (Natural England 2022k). Any habitat use in translocation should be capable of supporting a viable set of populations in a habitat (probably new, and developed ahead of removal (Natural England 2022k)). Such areas would need to be analogous to the removal areas, and at least as large and safe from damage or loss in the long term. It would need to be managed as part of a long-term, funded plan.

Species: protected plants, fungi and lichens	
Licensable activities	For the nine European Protected species, licences are required to handle or otherwise affect them (Natural England 2022k). These need securing in advance. If there is need to move species under licence, this will be as part of an agreed mitigation plan. Licences can only be given where it is clear that there are justifiable measures for affecting European Protected Species. Any habitat alteration will require a licence. Natural England will only issue a license if the proposal enhances European Protected Species habitats and helps to conserve them.
Monitoring	If mitigation is planned as part of a long-term funded programme, then it normally requires monitoring. Objectives will need to be stated. Timing will meet normal survey methods and frequency, and remediation actions undertaken as a result should be in line with plan requirements.
Summary	A PEA is required, with field and desk data. Times for field visits vary with species. A minimum of one visit is required, and subsequent monitoring should be spread over two years. Survey methods vary with species. Surveys are affected by weather, time of year, surveyor experience and scale. Wider coverage than the site alone is needed to fully understand potential population impacts. Mitigation needs to assess habitat impacts, and if unavoidable, will require site management. Any translocation requires pre-prepared suitable habitats and with long-term security and funded plans in place. Mitigation efficacy varies with treatment type.

PPFL = protected plants, fungi and lichens; BAP = Biodiversity Action Plan; PEA = Preliminary Ecological Appraisal; ZoI = zone of influence; NE = Natural England

Table 5.13 Species guidance: wild birds

Species: wild birds	
Site name and status	WB occur in designated and non-designated habitats across the UK. In England, site status can be assessed from Natural England's MAGIC website and is a basic part of a PEA desk search.
ZoI	WB occur in a range of environments and may use a series of different areas as part of their functional home range. This may well mean that the ZoI of a site for some species in practical terms is in excess of the 2 km zone used by desk searches.

(continued)

Table 5.13 Species guidance: wild birds (*continued*)

Species: wild birds	
Identified in pre-application discussions	Pre-application discussions will normally include all aspects of biodiversity, especially wild birds, which are one of the best documented species groups. Most LPA discussions will refer to validation checks and these will include WB in a number of areas. If potentially important WB sites are involved, and they are not in the discussion, this should be queried by LPAs.
The species should be picked up in the PEA	The PEA will include desk survey data sought from Local Record Centres, the National Biodiversity Network (NBN) and specialist groups as well as field surveys. The walkover element of a PEA will indicate suitable habitats and possible evidence of occurrence of particular species of WB. Absence in brief searches during a PEA visit would not be definitive due to the levels of effort, timing issues and weather issues related to detailed searches. Because of the rarity and protected status of some WB, standard desk searches may well draw some blanks, so that specialist groups or the specialists in the statutory agencies may need approaching where it is thought some WB might occur or where updates for recent status are sought. NE (2022) cautions against using Local Record Centre data that are older than two years. Field level data should be no more than two years old (CIEEM 2019). Absence from a desk search or lack of initial site evidence is not the same as site absence, and it may just be that there are no survey data for that site (NE 2022). Data older than two years need to be used with caution. TIN069 (NE 2010) notes the problems with Local Record Centre data and their use for projects such as windfarms. Note that WB are covered under a range of designations: – Wildlife and Countryside Act 1981: all WB are protected – Wildlife and Countryside Act 1981 schedule 1: 90 species listed under schedule 1.1 and 3 under schedule 1.2 – NERC Act 2006 S41: 49 species.
Protected species in and around the site	After using the PEA desk search as a screening for records in and around the site, walked coverage should extend beyond site boundaries as noted in the ZoI.
Age of data attached to the application	Field data are out of date within two years (CIEEM 2019b).
Objectives of surveys	The objective of the survey will determine the methods used and the level of effort, which will also be affected by the time of year and climatic variables. Objectives will include detailed surveys, while others may do an initial survey prior to more detailed work.

Species: wild birds		
		Objectives will include:
– presence/absence – enough to stimulate a detailed survey, or confirm basic status
– assessing which areas on a site are used by WB
– assessing site use in different parts of the year
– assessing species population sizes
– assessing potential habitat loss and management effects
– assessing potential effects of removing populations in case of partial/total area destruction
– potential effects of windfarms and associated developments.
Subsequent to any development, a round of monitoring surveys would be expected, with these matching techniques pre-development. These would be as part of a mitigation plan (Natural England 2022l). For many sites there may well be important cumulative effects to consider.
	Field collection methods	There is a range of methods available for WB, depending on the season and purpose of data collection.
Establishing baseline surveys are an important step for assessing impacts and for subsequent monitoring. Accordingly, it is important that methods are stated, along with their limitations and assumptions (BS 42020: BSI 2013).
Because of the range of species covered, specialists will be needed to carry out the surveys. Many species will have their own specialised survey methods, or group survey methods (Gilbert et al. 1998).
– Breeding bird survey – designed to assess the species breeding in an area. It provides numerical and spatial data
– Wintering bird survey – designed to establish the species using an area across winter
– Waterfowl counts – designed to understand the use of areas and numbers of waterfowl from autumn migration through to spring migration
– Vantage point surveys – designed to assess flightlines
– Windfarm assessments – designed to determine ground and air volume use in case of a planned windfarm development
– Species-specific methods – normally for breeding season, but some wintering species also have their own methods.

(continued)

Table 5.13 Species guidance: wild birds (*continued*)

Species: wild birds	
	1. Common Bird Census Breeding bird survey (CBC) Undertaken by a single observer mapping the use of an area between late March and July inclusive, recording all birds seen and heard in a standard notation (Bibby et al. 2000; Gilbert et al. 1998), it is possible to establish breeding bird territories and areas used/unused by each species. Gilbert et al. (1998) state that 10 visits are needed – at least 10 days apart – with 8–9 in the early morning an hour after dawn and 1–2 in the evening to catch those species that are more vocal then. Site visits by the same observer require that all areas are within 50 m of a recorder, and on farmland all hedges need to be covered. The same route is walked in alternate directions on subsequent visits to limit directional recording bias, and all birds seen or heard are noted. Analysis of summary maps of individual records allows identification of clusters, which are taken to be territories. Confirmatory behaviours such as carrying food or faecal sacs can be helpful. Single visits take 3–4 hours; longer introduces time of day bias on records and detection. This effectively places a limit of 10–20 ha for woodland and 50–100 ha for farmland. With fewer visits, either spaced more widely, or missing periods early or late in the season, there is risk of missing species or affecting the number of records that in turn affects analysis of territorial clusters. Bird maps are accompanied by a detailed habitat map, so that patterns of occurrence can be linked to habitats/crop practices as part of the impact assessment process. 2. Winter bird survey (WBS) The winter bird survey is used to map species in winter on transects and, like the CBC, used to link to habitats present. The most recent instructions (https://www.bto.org/sites/default/files/ewbs_instructions_2018.pdf) require monthly visits between December and March. If fewer visits are made, these must be in January and March. In order to reduce time effects, transects are walked at least one hour after sunrise and completed one hour before sunset. Records are marked on transect lines 100 m either side of the recorder, with sub-sections of 0–25 m, 25–100 m and beyond 100 m. Like other surveys, weather conditions are recorded, as are habitat and cropping types and stages. If using standardised surveys, the 1 km transect lengths are sub-divided into 100 blocks to readily link to habitat types for statistical analysis of habitat associations. The basic method is adaptable for development sites.

Species: wild birds

3. Waterfowl counts (WFC)

Waterfowl counts for development purposes are based on the standard UK Wetland Bird Survey (WeBS) count methodology. This includes wildfowl (ducks, geese and swans), waders, rails, divers, grebes, cormorants and herons. Gulls and terns are optionally included. In a typical year, over 220 waterbird species, races or populations are counted in UK WeBS counts https://www.bto.org/our-science/projects/wetland-bird-survey/about.

The basic principles of WeBS methods are counting on predetermined dates (daylight spring hightides). Each site is divided into sectors countable by a single person in less than four hours, with boundaries recognisable for subsequent repeat counts. These are based on clear vantage points, so that few birds are missed. Special emphasis is placed on water edges. Vantage points are visited using a recorded route. Gilbert et al. (1998) note the limitation of many binoculars as 500 m, and confounding features of weather, poor visibility and disturbance and error estimates. Count methods vary with flock sizes, and under such circumstances species identification will also become a problem.

In very cold periods waterfowl numbers may rise as birds come in from other areas in the short term; separate counts are required in these circumstances, in addition to the background counts. Where sites are used purely for overnight roosts, such as by some geese, mallards and teal, additional counts later in the evening will indicate site importance. These counts use the same methodology, but record their own times.

Geese, swans and low tide counts have their own methodologies (Gilbert et al. 1998).

4. Vantage point surveys (VPS)

Most survey methods are focused on birds on the ground, or on site. In some locations, such as potential windfarms, there is need to observe how birds use the air volume above the site, in order to understand the potential impact of turbines on flight patterns and mortality risk. Similarly, understanding how birds move in and out of sites where there is a large amount of interchange, such as wetlands over the course of a season or year, can be achieved by overviewing the site from a vantage point (VP). VPS can also be used to focus on breeding locations of endangered species or those prone to disturbance.

In order to understand site use, a series of non-overlapping VP 'viewsheds' are established, covering the site, and attempting to limit areas that are obscured and would potentially hide species and lead to under-sampling. Using the standardised recording methods established for windfarm assessments (SNH 2017), the height zones, time in zones and movements of individual birds are recorded on forms and mapped.

(continued)

Table 5.13 Species guidance: wild birds (*continued*)

Species: wild birds	
	Most VPS work outside of windfarms is non-standard, working areas for limited periods and with a wide range of objectives. Normally VPS limit the area covered to 2 km from the vantage point, beyond which species become increasingly hard to detect, and also limit survey durations to three hours or fewer to accommodate for fatigue. Additional time is allowed after reaching a cryptic VPS site to offset initial disturbance effects before commencing recording. As in other methods, detailed habitat maps are produced and subject to analysis for habitat preferences. Mapped and tabulated data are assessed to understand patterns of flights and site use for potential impact assessment. 5. Windfarm assessments (WFA) WFA require year-round data. Very small turbine proposals may need limited survey coverage (see SNH 2017). WFA require coverage of target species – primarily red list, schedule 1, NERC 2006 S41 and European Protected Species. Their year-round and within-day activity patterns need to be known. For large (>50 MW) proposals an additional 'control' area will also need surveying to act as a comparison. Depending on the habitats on the site, a range of survey methods will be required, possibly including moorland birds surveys (Brown and Shepherd 1993; Calladine et al. (2009) undertaken between mid-April to mid-July. For raptors and Short-Eared Owls, species-specific methods are required (Hardey et al. 2013). SNH (2017) sets out additional references for species/groups possibly found on site. Both the site and a 500 m band beyond are to be surveyed, with up to 6 km depending on species occurring (SNH 2017). For coastal species, a 2 km survey band is required. A lot of the year-round work involves VPS. Given detection problems increasing with distance, a limit of 2 km from individual VPs is recommended, with <1 km preferred (SNH 2017). Each VP has a scan of 180° so that several VPS will be required for a large site. VP watches take place across the day, limited to <3-hour sessions to limit fatigue, with 30-minute gaps between recording sessions. Areas observed should not be subject to disturbance, and require a minimum of 36 hours in breeding and non-breeding periods, making at least 72 hours spread across the year. Anything less needs to be justified and agreed. On a VP watch, each bird is plotted on a map and its height flown recorded for the period on standard forms when it is observed. These data are then used to estimate potential impacts of planned turbine placements and heights. SN (2017) specifies particular VP recording requirements according to species. 6. Species-specific methods (SSM). Details of SSM are available in Gilbert et al. (1998) and Hardey et al. (2013). For WFA, details are also provided in SNH (2017).

SURVEYS FOR PROTECTED SPECIES: WHAT THE LPA MIGHT HAVE ORDERED | 173

Species: wild birds	
Survey timing	1. CBC: March to July, with the start depending on position in UK and also expected species. Some species (e.g., willow tit, mistle thrush) will be most active before CBC starts (CIEEM 2020) and underestimated as a result. 2. WBS: December to March. 3. WFC. Formal WeBS counts cover the period September to March, with January being the key month. The full overwinter period gives a better impression of the seasonal use of the site. 4. VPS: depending upon the objective, VPS can be used across the year. Each period/season will have its own limiting factors, and these should be explicitly recognised. 5. WFA: for large areas, VP and other survey methods will have particular requirements. VPS occur across the year. 6. SSM: details vary with species.
Timing of visits in the day	1. CBC: avoid first hour after dawn. Registrations decline towards mid-day for most species, with early evening for later in the day as behaviours become more apparent after an earlier lull. For most locations, registrations drop by mid-July as breeding has occurred and territorial defence drops (Gilbert et al. 1998). 2. WBS: all surveys start at least an hour after dawn, finishing an hour before sunset. 3. WFC: ideally for tidal areas or near-coastal areas, within two hours either side of high tide. Inland sites: in less than four hours in the morning. 4. VPS: depending on the purpose, VPS use can cross most of the day. Periods of poor light at either end of the day can seriously affect detectability and reliability of data collected (SNH 2017). 5. WFA: where VP data are collected, these are meant to represent the risk pattern of day-round use according to species. For other survey objectives there will be specific time limits. 6. SSM: details vary with species.
Visit frequency	1. CBC: ideally 10 visits spread evenly across the study period. 2. WBS: four monthly individual visits. 3. WFC: one count per month. 4. VPS: according to objectives, survey frequency may vary, but should be mindful of the limits of small sample sizes and weather and other limiting factors that may skew extrapolation and data reliability. 5. WFA: visit frequency will depend on survey method. For VP surveys, no more than 9 hours can be worked within a 24-hour cycle, so that visits are spread across the seasons. 6. SSM: details vary with species.

(continued)

Table 5.13 Species guidance: wild birds (*continued*)

Species: wild birds	
Survey duration	1. CBC: 3–4 hours per visit. 2. WBS: Individual 1 km transect lengths take around 45 minutes. 3. WFC: two hours either side of high tide; less than four hours. 4. VPS: a maximum of three hours without a break (SNH 2017). 5. WFA: a maximum of three hours unbroken recording. 6. SSM: details vary with species.
Weather	1. CBC: detection is highly affected by wind and rain and heat/cold. 2. WBS: detection is affected by wind, rain, visibility and temperature. 3. WFC: strong winds, low cloud, rain/or poor visibility will strongly bias data. 4. VPS is strongly affected by wind, rain, temperature and visibility. 5. WFA: VP detectability depends on good ground visibility (>2 km) and a non-low cloud base. Strong winds, rain or snow/mist affect bird flight behaviour. 6. SSM: As (1).
Light	1. CBC: avoid first hour after dawn and poor light towards sunset; light conditions are affected by weather/cloud cover. 2. WBS: avoid first hour after dawn and poor light towards sunset; light conditions are affected by weather/cloud cover. 3. WFC: good light conditions are critical for both counting and identification. 4. VPS shares many of the limits noted for WFC. 5. WFA: See 3–4 above. 6. SSM: details vary with species.
Distance	1. CBC: all areas covered should be within 50 m of a field observer, and all areas of a site should be covered. This would include areas outside of the site as birds use a range of resources across the summer. 2. WBS: records are made for each 100 m either side of the transect, and beyond. 3. WFC: for effective counting within 500 m, but under good conditions some observers using telescopes can count much further but with error risk increasing (see windfarm limits for comparison). 4. VPS: depending on objectives, VP distance can be up to 2 km. 5. WFA: there is a limit of 2 km for detectability, with a preference for <1 km. 6. SSM: details vary with species.

Species: wild birds	
How far beyond development boundaries	1. CBC: include a minimum of 50 m from a boundary.
2. WBS: if the transect band goes beyond the site boundary, then records include all of these, plus those noted beyond.
3. WFC: at least 500 m to help understand the use of a site if it is part of a larger continuum.
4. VPS: depending on objectives, data collection should potentially extend 500 m beyond site boundaries to understand source and direction of flight lines.
5. WFA: at least 500 m for all components of the site and linked potential infrastructure lay-out.
6. SSM: details vary with species.
Years of data	1. CBC: at least one year's worth of visits to establish status in principle, mindful of the patchiness of temporal occurrence and possible error.
2. WBS: at least one year's worth of visits to establish status. If the area is subject to changes in cropping or usage, then two or more years would be suitable.
3. WFC: at least one year's data if not on a WeBS area. WeBS data provide a backdrop of trend data for comparison and context.
4. VPS. SNH (2017) recommends two years of data to allow for single sample variation and inter-year effects. If fewer data are collected, then reasons why, and implications need to be clearly stated and validated.
5. WFA: at least two years of field data are required.
6. SSM: depends on survey purpose.
Limitations	1. CBC; severely affected by time in year, time of day, weather, number of visits and recorder skills. CIEEM (2020) emphasised the impact of lost periods of time across the recording season on species detected, areas used, and conclusions drawn. In addition, Bibby et al. (2000) noted problems related to bird density, habitat and recorder speed and experience.
2. WBS: is affected by time of year – especially early or late passage times, weather, number of visits, distance – as birds smaller than thrushes are hard to locate with reliability at distance (Gillings et al. 2008), light and surveyor experience.
3. WFC: counts are affected by distance, weather and visibility, terrain, flock numbers, species composition, surveyor experience, disturbance, tidal state and more distant conditions.
4. VPS: counts are affected by distance, weather and visibility, terrain, flock numbers, species composition, surveyor experience, disturbance, distance, observer fatigue and number of visits and their time of year.
5. WFA: because WFA will often involve aspects of 1–4 methods, all of the above apply.
6. SSM: details vary with species.

(continued)

Table 5.13 Species guidance: wild birds (*continued*)

Species: wild birds	
Suitable licensed and experienced surveyors	1. CBC. There are no formal competency CIEEM requirements, nor are licences required for CBC surveys. As much of the CBC requires use of visual and auditory cues, accomplished levels of field skills are needed. 2. WBS: As for CBC, high levels of field skills are needed: both in bird identification and also habitat recording. 3. WFC: this requires highly skilled and experienced field surveyors: both in identification and in counting. Even the best surveyor will only get to within 90% of true total in estimating big flocks, whilst small flocks are overestimated (Gilbert et al. 1998). 4. VPS: requirements are shared with WFC. 5. WFA: because areas may support Schedule 1 protected bird species, due to the risk of disturbance whilst doing ground-based- rather than VP surveys- all surveyors will need to be licensed before commencing work. With the range of species and distance detection involved, all surveyors need to be highly experienced (SNH 2017). 6. SSM: details vary with species.
Determining impacts	Assessing impacts will require knowing the potential extent and form of impact and an indication of the species using part of or the whole area. Impacts will include: – habitat loss and deterioration – fragmentation of habitats or closing dispersal routes between habitat blocks – incidental mortality – management regime changes – soil water status changes – direct or secondary pollution – floodlighting areas within 50 m of woodlands, wetlands – disturbance from recreation, house building, infrastructure or windfarms – disturbance to large gardens, trees, veteran or large trees or other nesting sites. NB: as time will have elapsed from initial surveys, pre-construction surveys will be recommended for all areas thought to contain or be likely to contain protected species and take place <3 months ahead of planned commencement.

Species: wild birds	
Mitigating properly assessed potential impacts	Avoidance of any mechanical impact or indirect impact would be the very basic requirement for WB retention. Suitable avoidance actions (NS 2020a; Natural England 2022l) would include: – retaining and maintaining current habitats and protected species in situ – changing methods of working to avoid the active season for WB populations, especially breeding seasons – temporary fencing to control access to work areas – avoiding introduction of alien plants – increasing connectivity to suitable habitat areas, with clear evidence of suitability for each species involved – managing areas within the season to make them suitable as part of a coherent habitat management package – removing habitat features not used by birds before the breeding season, or displacement works before breeding might occur – proposals should include replacement nest site options and suitable habitat additions. Natural England (2022l) recommends: – no net loss of habitat – like-for-like replacement of nest sites near former nesting areas – better habitat than that lost – maintained habitat connections. Any habitat use in translocation should be capable of supporting a viable set of populations in a habitat (probably new, and developed ahead of removal (Natural England 2022l)). Such areas would need to be analogous to the removal areas and at least as large and safe from damage or loss in the long term. It would need to be managed as part of a long-term, funded plan.
Licensable activities	For the schedule 1 species, licences are required if there is a risk of disturbance to schedule 1 species, especially Barn Owls. Natural England cannot issue WB licences to allow development (Natural England 2022l).
Monitoring	If mitigation is planned as part of a long-term, funded programme, then it normally requires monitoring. Objectives will need to be stated. Timing will meet normal survey methods and frequency, and remediation actions undertaken as a result should be in line with plan requirements.

(continued)

Table 5.13 Species guidance: wild birds (*continued*)

Species: wild birds	
Summary	A PEA is required, with field and desk data. Times for field visits vary with species. Reliability of desk data is uncertain. At least one year of visits is required, and maybe more depending on development type, such as windfarms which require a minimum of two years. Survey methods vary with species and objectives/development type. Surveys are affected by a range of factors including weather, time of year, time of day, surveyor experience, scale, visibility, distance and fatigue. Limitations are critically important in assessing data reliability. Wider coverage than the site alone is needed to fully understand potential population impacts. Mitigation needs to assess habitat impacts, and if unavoidable, will require site management. Any translocation requires pre-prepared suitable habitats and with long-term security and funded plans in place.

WB = wild birds; PEA = Preliminary Ecological Appraisal; ZoI = zone of influence; NE = Natural England; CBC = Common Bird Census Breeding bird survey; WeBS = Wetland Bird Survey; VPS = Vantage Point Surveys; VP = vantage point; WFA = windfarm assessments; SSM = Species-specific methods

Table 5.14 Species guidance: invertebrates

Species: invertebrates	
Site name and status	Invertebrates (IV) occur in designated and non-designated habitats across the UK. In England, site status can be assessed from NE's MAGIC website and is a basic part of a PEA desk search.
ZoI	IV occur in a range of environments, and some species may use a series of different areas as part of their functional home range of their life stages. This may well mean that the ZoI of a site for some species in practical terms is in excess of the 2 km zone used by desk searches. That would also occur where some sites work as part of a super site, connected and exchanging components as one site declines while another flourishes.
Identified in pre-application discussions	Pre-application discussions will normally include all aspects of biodiversity, including invertebrates. Most LPA discussions will refer to validation checks and these will include IV in a number of areas. If potentially important IV sites are involved, and they are not in the discussion, this should be queried by LPAs.

Species: invertebrates	
The species should be picked up in the PEA	The PEA will include desk survey data sought from Local Record Centres, the National Biodiversity Network and specialist groups as well as field surveys. The walkover element of a PEA will indicate suitable habitats and possible evidence of occurrence of particular species of IV. Absence in brief searches during a PEA visit would not be definitive, due to the levels of effort and timing issues, and weather issues related to detailed searches.
CIEEM (2020) noted that the presence of priority habitats, mosaics of high value habitat in or nearby the site, veteran trees and connectivity to habitats are all indicators of potential invertebrate interest, in addition to a site's nature conservation designation. If a site has a nature conservation designation, CIEEM suggested that it might deserve a detailed phase 2 Ecological Impact Assessment (EcIA) invertebrate evaluation.
Because of the rarity and protected status of some IV, standard desk searches may draw some blanks, so that specialist groups or the specialists in the statutory agencies may need approaching where it is thought some IV might occur, or updates for recent status are sought.
Field level data should be no more than two years old (CIEEM 2019b; Natural England 2022m). Absence from a desk search or lack of initial site evidence is not the same as site absence and it may just be that there are no survey data for that site (Natural England 2022m). Data older than two years need to be used with caution.
Note that IV are covered under a range of designations:
– Wildlife and Countryside Act 1981 schedule 5: 399 species listed under schedule 5
– NERC Act 2006 S41: 3 species.
Protected species in and around the site	After using the PEA desk search as a screening for records in and around the site, walked coverage should extend beyond site boundaries as noted in the ZoI.
Age of data attached to the application	Field data are out of date within two years (CIEEM 2019b).
Objectives of surveys	The objective of the survey will determine the methods used and the level of effort, which will also be affected by the time of year and climatic variables (Drake et al. 2007). Objectives will include:
– presence/absence – enough to stimulate a detailed survey, or confirm basic status
– assessing which areas on a site are used by particular IV
– assessing site use in different parts of the year
– assessing known species population sizes
– assessing potential habitat loss and management effects on IV
– assessing potential effects of removing populations in case of partial/total area destruction.

(continued)

Table 5.14 Species guidance: invertebrates (*continued*)

Species: invertebrates	
	Subsequent to any development, a round of monitoring surveys would be expected, with these matching techniques pre-development. These would be as part of a mitigation plan (NE 2022). For many sites, there may well be important cumulative effects to consider.
Field collection methods	There are a range of methods available for IV, depending on the season and purpose of data collection and species group (Drake et al. 2007), but there is no simple guidance available to inform whether terrestrial or aquatic invertebrate surveys are required, nor the extent of suitable surveys (CIEEM 2020). NE (2022) recommends surveys for high value areas such as woodland with ancient trees, semi-natural vegetation, wetlands, coastal areas and habitat mosaics on previously developed land. Establishing baseline surveys are an important step for assessing impacts and for subsequent monitoring. Accordingly, it is important that methods are stated, along with their limitations and assumptions (BS 42020: BSI 2013). Because of the range of species, and environments, covered, specialists will be needed to carry out the surveys. Many species will have their own specialised survey methods, or group survey methods (Drake et al. 2007). Methods include sweep nets, beating, pitfall traps, active searches, light traps, standard timed transect lines, vacuum sampling and pond netting. Given the variation in species emergence times, basic requirements to begin to understand a site's IV include: – five visits late April to September – more targeted visits dependent on PEA results or particular habitat types (NE 2005). Wetlands tend to have different timetables due to their distinctive communities, and many species have limited periods (such as bees) when they are on the wing. Miss these and the results are problematic. As a rough guide (NE 2005, Buglife 2019) an average site of 10–50 ha should have 3–7 days of field work, 3–7 days of identification and 2–5 days of report writing. The time varies with breadth of the groups covered. Note that protocols will vary between terrestrial and aquatic communities. Drake et al. (2007) provide outlines for sampling target groups on a site. The key here is to determine in advance which groups are to be sampled, and why. Each method has its strengths and limitations, and these need to be stated and understood when reports are evaluated.

Species: invertebrates	
Survey timing	April to October, but individual groups/species will have focal times within or beyond this and may be sampled over a longer period (Drake et al. 2007). Four to five visits will be the minimum. Some species, such as Fairy Shrimp *Chirocephalus diaphanus*, exploit autumnal ephemeral pools and their whole lifecycle is over in three weeks, making timing of surveys critical.
Timing of visits in the day	Activity periods for different groups vary across a 24-hour period, so timings will be determined by target species or groups.
Visit frequency	5–7 visits across the course of the season, supplemented or replaced according to target groups/species.
Survey duration	Duration varies with methods and sites.
Weather	Warm, dry, low wind speeds; avoid previous rain periods. Cold periods depress IV activity, and prolonged dry periods negatively influence IV recording. Within the day, very hot periods collect fewer IV than later in the day (Drake et al. 2007).
Light	Light conditions and their effects vary across target species and groups.
Distance	All areas of the site should be sampled, according to the habitat types, with habitats 50 m beyond evaluated. NE (2005) recommends checking up and down stream in aquatic systems, but no distances are given.
How far beyond development boundaries	As communities are usually linked, 50 m is the minimum.
Years of data	At least one year's worth of data is required.
Limitations	IV surveys are affected by a wide range of issues, including: light, weather, season, temperature, survey timing, preceding weather conditions, surveyor experience, number and spread of visits across the active season, and severely affected by time in year, time of day, weather, number of visits and recorder skills.
Suitable licensed and experienced surveyors	There are no formal competency CIEEM requirements for IV, although there is a competency listing for aquatic invertebrates. Competencies for White-clawed Crayfish and Pearl Mussels are covered elsewhere. Drake et al. (2007) and Buglife (2019) advocate the use of very experienced surveyors, otherwise results can be seriously skewed. European Protected Species licences are required for Large Blue butterfly *Phengaris arion*, Fisher's Estuarine Moth *Gortyna borelii lunata* and Little Ramshorn Whirlpool Snail *Anisus vorticulus* where there is risk of disturbance (Natural England 2022m).

(continued)

Table 5.14 Species guidance: invertebrates (*continued*)

Species: invertebrates	
Determining impacts	Assessing impacts will require knowing the potential extent and form of impact and an indication of the species using part of or the whole area. Natural England (2022m) suggests putting species data collected on surveys into the Pantheon database (https://pantheon.brc.ac.uk/) to help assess the potential value and risks to that site from development. This underlines the need for reliable data sets. Impacts will include: – habitat loss and deterioration – fragmentation of habitats or closing dispersal routes between habitat blocks – incidental mortality – management regime changes – soil water status changes – water flow or status changes – direct or secondary pollution – floodlighting areas within 50 m of woodlands, wetlands – disturbance from recreation, house building and infrastructure – disturbance to large gardens, trees, veteran or large trees or other nesting sites. NB: as time will have elapsed from initial surveys, pre-construction surveys will be recommended for all areas thought to contain or be likely to contain protected species and take place <3 months ahead of planned commencement – mindful of the limitations applied to the IV concerned.
Mitigating properly assessed potential impacts	Avoidance of any mechanical impact or indirect impact would be the very basic requirement for IV retention. Suitable avoidance actions (Natural England 2022m) would include: – retaining and maintaining current habitats and protected species in situ – changing methods of working to avoid the active season for IV populations, especially breeding seasons – temporary fencing to control access to work areas – increasing connectivity to suitable habitat areas, with clear evidence of suitability for each species involved – managing areas within the season to make them suitable as part of a coherent habitat management package – proposals should include replacement and suitable habitat additions. Natural England (2022m) recommends: – no net loss of habitat – better habitat than that lost – maintained habitat connections.

Species: invertebrates	
	Any habitat use in translocation should be capable of supporting a viable set of populations in a habitat (probably new and developed ahead of removal (NE 2022)). Such areas would need to be analogous to the removal areas, and at least as large and safe from damage or loss in the long term. It would need to be managed as part of a long-term, funded plan.
Licensable activities	For the three European Protected Species, licences are required if there is a risk of disturbance.
Monitoring	If mitigation is planned as part of a long-term, funded programme, then it normally requires monitoring. Objectives will need to be stated. Timing will meet normal survey methods and frequency, and remediation actions undertaken as a result should be in line with plan requirements.
Summary	A PEA is required, with field and desk data. Times for field visits vary with species. Reliability of desk data is uncertain. A minimum of one year of survey visits is required. Survey methods vary with species and objectives/development type. Surveys are affected by a range of factors including weather, time of year, time of day, surveyor experience, scale, visibility, distance and temperature, visit timing and number of visits and season. Wider coverage than the site alone is needed to fully understand potential population impacts. Mitigation needs to assess habitat impacts, and if unavoidable, will require site management. Any translocation requires pre-prepared suitable habitats and with long-term security and funded plans in place. Efficacy of mitigation options is uncertain.

PEA = Preliminary Ecological Appraisal; ZoI = zone of influence; NE = Natural England; IV = invertebrates

Table 5.15 Species guidance: bats

Species: bats	
	Bats are protected under the Habitats and Species Directive.
Site name and status	Bats occur in designated and non-designated habitats across the UK. In England, site status can be assessed from Natural England's MAGIC website and is a basic part of a PEA desk search.
ZoI	Bats occur in a range of environments across the course of the year, and in different areas, as part of their functional home range. The ZoI of a site is up to 10 km for desk search purposes for high-risk species, and 5 km for all other bat species (Collins 2016). The desk search should extend to 10 km for sites designated as special areas of conservation or sites of special scientific interest for bats. For major infrastructure developments, 1 km is required for surveys (Collins 2016).
Identified in pre-application discussions	Pre-application discussions will normally include all aspects of biodiversity, including bats. Most LPA discussions will refer to bats. If bat sites are involved, and they are not in the discussion, this should be queried by LPAs.
The species should be picked up in the PEA	The PEA will include desk survey data sought from Local Record Centres, the National Biodiversity Network and specialist groups as well as field surveys. The walkover element of a PEA will indicate suitable habitats and possible evidence of occurrence of bats. Absence in brief searches during a PEA visit would not be definitive, due to the levels of effort and timing issues, and time of year. A PEA walkover may note suitable sites for bats, but most roosts will be missed and will require a phase 2 evaluation. Bats are covered under a range of designations: – Conservation of Habitats and Species Regulations (2017): all bat species. – Wildlife and Countryside Act 1981: all bat species. – NERC Act 2006 S41: Barbastelle, Bechstein's, Brown Long-eared Bat, Greater and Lesser Horseshoe Bats, Noctule and Soprano Pipistrelle.
Protected species in and around the site	After using the PEA desk search as a screening for records in and around the site, walked coverage should extend 500 m beyond site boundaries. It may extend to 1,000 m for large infrastructure projects (Collins 2016).
Age of data attached to the application	Field data are out of date within two years (CIEEM 2019b; NS 2020a).

Species: bats	
Objectives of surveys	Because of their European Protected Species status, any development that risks disturbing bats needs to be sure of the potential impacts. Objectives will include: – assessing whether the site (building/tree/set of trees/larger area) has the potential to support bats (potential roost features) – identify if potential roost features occur – assessing which areas on a site are used by bats, and by which species: for foraging, commuting and breeding/winter/seasonal roosting – assessing bat population sizes – assessing potential habitat loss and management effects on bats – assessing potential effects of removing habitats and/or roost site in case of partial/total area destruction. Subsequent to any development, a round of monitoring surveys would be expected, with these matching techniques pre-development. These would be as part of a mitigation plan (NE 2022). For many sites, there may well be important cumulative effects to consider.
Field collection methods	Methods depend on the site and potential development type. The objective is to determine whether or not there is likely to be an impact. This may be direct through roost habitat alteration, or indirect through changes to areas used by bats, or novel causes of potential mortality such as turbines. Each will demand their own set of survey protocols. The level of input will also vary with habitat suitability (Collins 2016). The first issue is to assess whether there are likely to be bats in/around the risk feature. This should have been outlined by the PEA desk study, supplemented by a first visit to the site. Where the site involves buildings or trees, the initial job is to produce an assessment of habitat suitability. This will influence the number of visits and the methods to be used. Locations that have low suitability for bats will need lower levels of input than high suitability roosts, which can potentially be expected to have more bats. In potential impact terms, it is important to understand the ways in which data can be collected. There are 3 main ways: Visual: bats are extremely hard to observe in flight, and to identify even some of them by sight is difficult for all but the most experienced.

(*continued*)

Table 5.15 Species guidance: bats (*continued*)

Species: bats
As part of a preliminary roost assessment (PRA), external visual cues for possible roosts include signs of existing or recent roost use, such as droppings, oily marks on hole margins, lack of cobwebs across a suitable hole, or scratch marks, which are the best starters but not guaranteed to confirm bat status, and smaller roosts are especially likely to be missed by external inspections (Froidevaux et al. 2020). Most of the signs are shared by trees and buildings/natural features. Collins (2016) noted that failure to find signs of bats at a suitable roost site on a single visit – especially if not at the peak time (May to August) – is not sufficient to discount the site, while inspections at a site normally are. After an external PRA, internal visual PRAs by a suitably licenced ecologist will be needed for a building, looking for much the same signs of bat presence as outside. Inside, droppings, corpses and bats themselves may well be evident. Nooks and hidden areas will need searching by endoscope (a licence-requiring activity). Presence and absence of evidence is always to be recorded (Collins 2016). On a potential tree roost, similar licenced visual work will be needed. Collins (2016) noted that failure to find a tree roost on a single survey is not the same as confirming absence, as bats may move round frequently across the April to October activity period. If there is evidence/potential evidence of bats, then confirmation either way is sought. This will include some visual work but normally linked to some sort of bat detection technology. Visual observation of bats leaving roosts is part of a roost assessment process, but relies on support from bat detection recognition technology to confirm identity. Visual observation alone will miss 78% of bats as light fades (BCT 2022). For some very late-emerging species this is not an option. Similarly, it is possible to monitor bat returns to roost, but visual cues alone are inadequate (Collins 2016) and unreliable by themselves (BCT 2022). For both, night vision aids (NVA) are important. BCT (2022) recommends that as dusk surveys are as efficient as dawn surveys, the former, aided by NVA should replace dawn surveys. *Detection technology* Fixed and portable bat detectors are used to detect bats and identify bats at exit/entry from roosts, and as they use the surrounding countryside or urban areas. Suitable bat detectors, properly placed, are able to detect and record species and allow identification, and relative use, but not absolute numbers of bats. Using detectors with recording facilities allows remote recording and playback of data for identification and quantification purposes (Collins 2016). In addition to standard static or portable detectors, thermal imaging cameras are now recommended for roost emergence, and especially for roost return surveys (BCT 2022) where a combination of observer error and detection difficulty can be overcome – including playback allied to standard detectors. Infrared imagery can also be used in limited circumstances.

Species: bats	
	In windfarm proposals, a limited amount of pre-dusk vantage point work, viewing large, early emerging bats such as noctules may also be used. This is a very limited area of work.

3. eDNA

Like most other species (see GC Newts for example) it is possible to determine species identity without seeing the actual animal. Bat droppings can be used to assess which species may be involved. Experienced ecologists will be able to identify species by their droppings, but in many cases, this is difficult, hence the use of eDNA tests. Care should be taken not to mix samples where different locations or ages are concerned. eDNA tests will indicate which species have been present.

Assessing risks to bats
1) Desk and field assessment via a PEA.
2) Undertaking a PRA on a building/natural feature/tree.
3) If 1 or 2 suggest a bat risk, then a selection of activity, static or transect activity surveys, would need to be undertaken, depending on the PRA results, scale of the site and planned development and also the roost type. Hibernation roost work can only be undertaken in winter, while presence/absence surveys should be undertaken between mid-May and August (Collins 2016).
3.1) PRA are ranked as low, moderate and high (Collins 2016). Moderate and high require presence/absence surveys (Box 1).

Box 1. Survey needs according to roost potential

High roost potential	Moderate roost potential	Low roost potential
Three dusk emergence and/or pre-dawn re-entry surveys May to September	Two dusk emergence and/or pre-dawn re-entry surveys May to September	One dusk emergence and/or pre-dawn re-entry surveys May to September

Dusk emergence surveys start 15 minutes before sunset and last for two hours after sunset. Pipistrelle and serotine surveys start 30 minutes before sunset.
Pre-dawn re-entry surveys start 1.5 to 2 hours before sunrise and continue until at least sunrise. Note: BCT (2022) recommends use of dusk surveys only, aided by NVA. Optimum period May to August.

(continued) |

Table 5.15 Species guidance: bats *(continued)*

Species: bats	
	Emergence counts require several observers to cover all areas of a building/tree. Recording of data is required, and BCT (2022) recommends use of NVA.
	Linked to the re-entry surveys are back-tracking surveys, where a combination of visual tracking and bat detectors are used to try and follow bats back to a roost. BCT (2022) is pertinent here on efficacy.
	3.2) Roost characterisation survey (RCS)
	If 3.1 reveals a roost, then it is important that the LPA can make an informed judgement on possible impacts. An RCS is needed to understand impacts and mitigation options. Collins (2016) recommends that the following is given in an RCS:
	– number of bats in colony; size is critical, so multiple counts are needed to understand the size and how the roost is used
	– access points used: knowing these helps understand impacts and flight lines. Details of distances to vegetation are important for effective mitigation
	– temperature and humidity regime in roost: at least a season's data on regimes in case of need for optimal mitigation
	– aspect and roost orientation: these are needed in case of future mitigation
	– size and perching points: to be provided in case roost replication is required
	– lighting: developments typically involve light changes. Existing internal and external light regimes are to be provided.
	– surrounding habitat: what is available and how it connects departing bats to other habitats and feeding areas.
	3.3) Swarming surveys
	Some locations are used for breeding, when bats mass together over a short period in the August to October period (Box 2).

Box 2: Timing for activity surveys, including swarming surveys

Survey objective	Dusk survey	Pre-dawn survey
Bat activity away from roost transects	Start: 15 minutes before sunset. For: 2–3 hours	Length 1.5–2 hours. Finish sunrise
Mating activity and swarming sites	Start: at sunset. For: 4+ hours after sunset	–

| Species: bats | Swarming surveys normally also use static recorders, covering the period of darkness, and are also suitable for NVA use.

3.4) Static recording/automated survey

The ability to record the use of an area by bats over a night, or more typically a series of nights across the seasons, allows the building up of the profile of species and their frequency. Static recording, aka automated survey, uses remotely placed bat detectors set to open 30 minutes before sunset and close after three hours or at dawn. The data collected are then analysed for species identities and pattern and number of calls each night. Statics can be used to supplement visual or NVA recording at single locations and as part of a programme of understanding spatial use over a large area in support of transect recording (Collins 2016). Statics come with a series of limitations that need to be understood if used in support of planning applications.

Best undertaken between mid-May and mid-September, statics are normally run with data loggers that record temperature and humidity: both affect bat behaviours.

3.5) Transects

When an area is big enough to need multiple recording locations, it is also likely to contain a range of habitat types, including hedges. Such areas would not be effectively sampled by statics alone. Instead, these are best worked by walking transect lines. Transect lines cover all habitat types and wend across an area in a pre-determined way for 2–3 hours after sunset, so that activity and locations are recorded for later analysis. As transects are typically repeated across a season, and between seasons, direction walked is reversed each time to limit directional bias. Transects can be supplemented by use of statics – which expands the temporal coverage, or point counts at given locations when the observer stands and records for three minutes (Collins 2016).

3.6) Medium to large areas

Collins (2016) provided a summary of the mix of transects and statics that would provide the baseline with which to understand the potential effects of a development on bats.

(continued) |

Table 5.15 Species guidance: bats (*continued*)

Species: bats

Box 3. Frequency of visits for transects and static surveys

Size	Low habitat quality	Medium habitat quality	High habitat quality
<1 ha	T: 1 T twice March to September. Best: June to August	T: 1 T 3 March to September. Best: June to August. At least 1 T at dawn and dusk within 24 hr period	T: 1 T 4 March to September. Best: June to August. At least 1 T at dawn and dusk within 24 hr period
	S: 1 location twice for 3 consecutive nights March to September	S: 1 location three times for three consecutive nights March to September	S: S:1 location four times for three consecutive nights March to September
1–15 ha	T: 1 visit per season (spring, summer, autumn)	T: 1 T per month April to October. At least 1 T at dawn and dusk within 24 hr period	T: Up to 2 T per month April to October. At least 1 T at dawn and dusk within 24 hr period
	S: 1 location per T. Data on 3 consecutive nights per season	S: 1 location per T. Data on 3 consecutive nights each month April to October	S: 2 locations per T. Data on 4 consecutive nights each month April to October
>15 ha	T: 1 visit per season (spring, summer, autumn)	T: 1 T per month April to October. At least 1 T at dawn and dusk within 24 hr period	T: Up to 2 T per month April to October. At least 1 T at dawn and dusk within 24 hr period
	S: 1 location per T. Data on 4 consecutive nights per season	S: 2 location per T. Data on 5 consecutive nights each month April to October	S: 3 locations per T. Data on 5 consecutive nights each month April to October

T= transects; S= static surveys

Species: bats

3.7) Windfarms

Windfarms provide a series of possible impacts and a novel way of death: barotrauma where rapid internal pressure changes affect organs leading to death. As windfarms operate on a functional and spatial scale different from other developments, there is a separate stand-alone reference produced by the UK statutory conservation agencies (https://www.nature.scot/doc/bats-and-onshore-wind-turbines-survey-assessment-and-mitigation) (NS 2021). This is the accepted standard used by NE and the wind power industry. There are several stages to follow.

3.7.1 Desk study:
– collation of all records within 10 km of the proposed site
– location of all turbines (single or multiple) within 10 km to assess potential cumulative effects
– position of the proposal in species ranges; edge of range has more potential impact on high-risk species than in range core

3.7.2 Bat surveys:
these aim to:
– identify species using the potential site and their spatial and temporal distributions
– location of roosts/swarming areas in the surrounding area that could be affected by the site
– location and extent of foraging habitats on the site and nearby that might be affected by a development

3.7.3 Survey methods:
– Roosts: search for roosts/swarming site within 200 m + rotor radius
– Activity surveys: these are undertaken in conditions >10°C, wind speeds <5 m/sec and no rain, and weather data are recorded on site. Full spectrum static detectors are used, opening 30 minutes before sunset and closing 30 minutes after sunrise. See Box 4, overleaf.

(continued)

Table 5.15 Species guidance: bats *(continued)*

Species: bats

Box 4: Survey methodology for use on windfarm proposals

Ground level static surveys	10 nights consecutive recording (weather allowing) April to May; June to mid-August; mid-August to October	Placement of statics where turbines will be located. If >10 turbines, 10 locations + additional statics for one third of the number of turbines	If in woodland, nearest opening to proposed turbines
Statics at height	As ground, but only where high levels of activity at rotor-swept area, if existing mast or weather mast to be erected	Records beyond range of ground-based statics for key-hole sites	If in woodland, nearest opening to proposed turbines
Walked transects and vantage points	Transects: Discretionary; linked to static frequency. Linked to linear features and watercourses	Vantage points: used from overview areas to monitor early emerging bats in open areas	Follow protocols in Box 1 and Box 3.

3.6.4 Interpreting the data collected:
It is important to understand whether the site might be busier or typical for the wider area. NS (2020) provides a way of estimating this, and this should be in the material provided to the LPA.

Vulnerability: Tables in NS (2021) set out vulnerability of bat species to collision risk and their overall risk. These are then added to by looking at site-specific risk, based on mapped habitat assessments provided as part of the field survey work. A final overall risk assessment score is then presented. This allows an LPA to understand the risks, and also to see mitigation options as set out in NS 2021.

Species: bats

Survey timing	1. Desk and field PEA: all year round in principle. 2. PRA: all year round in principle, but after leaf fall trees are easier to assess. 3. Hibernation survey: October to March 4. Field survey: roost assessment: April to October 5. Roost Characterisation Survey: based on 3.1: field work occurs April to October 6. Swarming survey: August to November 7. Static survey: mid-May to mid-September 8. Transects: April to October; best mid-June to August 9. Large areas: April to October 10. Windfarms: April to October.
Timing of visits in the day	1. Desk and field PEA: daylight hours to assess habitats and possible roost options. 2. RPA: daylight hours 3. Hibernation survey: as these are normally dark/very poorly lit, day or night 4. Field survey: roost potential – daylight hours 5. RCS: If using activity data, a combination of sunset-sunrise records. As both vary over time, it is important that these times are stated, rather than implied. 6. Swarming survey: sunset for 4 hours 7. Static survey: 1/2 hour before sunset–sunrise 8. Transects: 15 mins pre sunset for 2–3 hours; pre-dawn 1.5–2 hours, finishing at sunrise 9. Large areas: as transects and static surveys 10. Windfarms: as transects and static surveys.
Visit frequency	1. Desk and field PEA: an initial visit leading, if risks are noted, to more detailed phase 2 survey work. 2. PRA: a visit in daylight hours. Over a large site this may take several days 3. Hibernation survey: a single visit 4. Field survey: roost potential: daylight hours 5. RCS: RCS is the summation of multiple visits collating static and transect and emergence work 6. Swarming survey: over a series of nights. No fixed frequency in Collins (2016) 7. Static survey: varies with roost type and development type and scale – see boxes 8. Transects: varies with roost type and development type and scale – see boxes 9. Large areas: as transects and static surveys 10. Windfarms: see Box 4.

(continued)

Table 5.15 Species guidance: bats (*continued*)

Species: bats

Survey duration	Duration varies with methods and sites: see boxes.
Weather	Best in warm, dry, low wind speed conditions; avoid rain, mist, strong winds (>5 m/s (NS 2021) and temperatures below 10°C at dusk. Poor conditions depress bat activity (Collins 2016, NS 2021).
Light	The emergence and return times of bats vary significantly (Collins 2016, Andrews and Pearson 2016) and can include relatively light or dark conditions.
Distance	10 km for desk searches, and at least 500 m for field surveys. 10 km for cumulative impacts for windfarms and possibly further with risks to swarming sites or flight lines (NS 2021). For possible impacts on bats using a site, a 50 m buffer from trees/hedges/wetlands is recommended (NS 2021).
How far beyond development boundaries	500 m is the basic radius, with 1 km for large developments (Collins 2016).
Years of data	Collins (2016) recommends at least two years of data for large sites, with one to three in others. At least one year's worth of data is required. NS (2021) requires new data when surveys are more than two years old. NS (2021) recommends at least three years of post-development monitoring data.
Limitations	Bat activity varies strongly within and between years (Collins 2016; NS 2021), so full seasons of activity data are required (NS 2021). Activity levels vary across the active year, so loss of any periods of data within these has potentially serious effects and should be stated. Weather effects are especially important. Activity areas are affected also by changes in cropping regimes across the year. Also, surveying large areas with inadequate numbers of static recorders will affect records and inference drawing (NS 2021). Starting surveys late will impact recording levels, and placement of statics far away from linear features will affect records, as detectability of bats varies significantly with distance (Adams et al. 2012). Roost determination is a major limitation (CIEEM 2021; Andrews 2018) on finding and determining roost size, type and likely effects. Interpreting bat recordings is a major factor in species identification and interpretation, and experience with both technology and software is potentially a major limitation. Field experience is also a limiting factor in detecting and interpreting bat records. The potential impacts or their lack of impacts needs to be explained and validated, rather than discounted without supporting material.

Species: bats

Suitable licensed and experienced surveyors	To work on bats and risking disturbance requires a level 2 bat licence or more. Roost entry and endoscope work both require licences, and trapping and handling or tracking requires a higher licence. CIEEM expects a high level of training in its competency document (CIEEM 2013). Where impacts are expected, a mitigation licence will be needed.
Determining impacts	Impacts will be direct: through roost destruction, direct mortality or indirectly through habitat, foraging route or linked changes. Understanding potential impacts requires a suitable body of reliable data collected within two or fewer years (NS 2021). LPAs need to understand the possible risks based on:<ul><li>distributional records: the impacts on individual species depend on the position of the species within its range, as well as its relative abundance (NS 2021).</li><li>historical records and data recency</li><li>the potential impact of the proposal on the roost type and internal roost geography.</li><li>the bat species composition and levels of use of the areas surveyed</li><li>potential cumulative effects of other proposals in the area</li><li>for windfarms, the potential impacts depend on the siting of turbines (>50 m from suitable habitats/linear features), species present, rotor sweep height and speed and pattern of turbine placement, as well as potential cumulative effects.</li></ul>
Mitigating properly assessed potential impacts	Mitigation options depend on the proposal, scale and type, as well as the individual requirements of the species potentially involved (Collins 2016, NS 2021):<ul><li>the simplest option is to not undertake the proposal at that site: avoidance. Failing that, alterations in methods and/or timing may reduce effects (Natural England 2022n)</li><li>start work outside of the maternity season: May to August</li><li>use bat-friendly chemicals, if they are to be used</li><li>if roosts are destroyed, there must be like-for-like roost creation within the building, or lighting alteration to allow access to foraging habitats and away from roost entrances/exits.</li><li>installation of bat boxes or a bat building may help</li><li>instating/improving access to and quality of habitats</li><li>for tree roosts, similar mitigation is required as in buildings</li></ul>

(continued)

Table 5.15 Species guidance: bats (*continued*)

Species: bats	
	• bats are impacted by road schemes. Where construction cannot be avoided, under or over passes must follow pre-existing routes. Otherwise, mortality will be expected as they follow 'ghost trails' (Berthinussen and Altringham 2012). Or deter bats from crossing using habitats and lighting. • for windfarms, mitigation includes: adjusting turbine layouts to avoid areas with high bat use, making 50 m+ buffers from important commuting lines and habitats, reducing idling speeds of blades, or raising the cut-in speed – matching the wind speeds when bat activity decreases – to reduce mortalities (NS 2021).
Licensable activities	Disturbance to a European Protected Species requires suitable bat species licence and a subsequent mitigation licence.
Monitoring	Monitoring is required as part of a plan and should clearly identify what needs to be monitored, when and how any results will be responded to (Natural England 2022n, NS 2021). Methods should mirror those pre-development to allow comparison. Windfarms require a minimum of three years of post-development monitoring. Carcase detection post-development is important to assess the predicted impacts of a scheme. Use of dogs is described in NS (2021) and has far higher success rates than trained human searchers.
Summary	A PEA is required, with field and desk data. Times for field visits vary with species and development types. A minimum of a full year of surveys is required. Survey methods vary with species and objectives/ development type. Surveys are affected by a range of factors including weather, time of year, time of day, surveyor experience, scale, visibility, distance and temperature, visit timing and number of visits and season. Wider coverage than the site alone is needed to fully understand potential population impacts. Up to 10km is needed in data searches. Mitigation needs to assess habitat impacts, and if unavoidable, will require site management. Mitigation efficacy is uncertain (Lintott and Mathews 2018). Any agreed effects on bats requires a detailed mitigation plan, pre-prepared suitable habitats and with long-term security and funded plans in place. Windfarms have a clear set of mitigation tools to be put in place.

PRA = preliminary roost assessment; NVA = night vision aids; RCS = roost characterisation survey

Table 5.16 Species guidance: Great Crested Newt

Species: Great Crested Newt

	Great Crested Newts are protected under the Habitats and Species Directive.
Site name and status	GCN occur in designated and non-designated habitats across the UK. In England, site status can be assessed from Natural England's MAGIC website and is a basic part of a PEA desk search. eDNA data for GCN are now available via MAGIC.
ZoI	GCN occur in a range of environments across the course of the year and different areas as part of their functional home range. The ZoI of a site is normally less than the 2 km zone used by desk searches. Some sites work as part of a super site, connected and exchanging components as one site declines while another flourishes. The standard search distance (Natural England 2022o) is 500 m.
Identified in pre-application discussions	Pre-application discussions will normally include all aspects of biodiversity, including GCN. Most LPA will refer to validation checks and these will include GCN. If GCN sites are involved, and they are not in the discussion, this should be queried by LPAs. As GCN may well be covered by district licences (DL), GCN should be an important element in a pre-application discussion.
The species should be picked up in the PEA	A PEA will include desk survey data sought from Local Record Centres, the National Biodiversity Network and specialist groups as well as field surveys. The walkover element of a PEA will indicate suitable habitats and possible evidence of occurrence of GCN. Absence in brief searches during a PEA visit would not be definitive, due to the levels of effort, timing issues and time of year. For GCN, a walkover would include determination of permanent/seasonal waterbodies within 500 m of the development, or suitable refuges such as log piles, rubble, hedgerows, grassland or scrub. If the waterbodies are thought suitable, then it is normal for a 10 factor Habitat Suitability Index score to be produced (ARG 2010). Optimal times for this are March to June inclusive. With values between 0 and 1, the waterbody is scored, with higher scores meaning the body is more likely to support GCN. Note that low scoring bodies can still support GCN.

(*continued*)

Table 5.16 Species guidance: Great Crested Newt (*continued*)

Species: Great Crested Newt

	Because GCN are European Protected Species, it is important to consider whether the site or immediate area support GCN. This can be done by detailed sampling or through DL. If the LPA is part of a DL scheme (Natural England 2022o), there will be maps of areas where GCN live and the most important areas to conserve. These risk areas are: – red: where DL does not apply. – amber: where there are GCN populations and habitats and dispersal routes. DL apply. – green: zones with areas with GCN. DL apply. Note that GCN are covered under a range of designations: – Conservation of Habitats and Species Regulations (2017) – Wildlife and Countryside Act 1981 – NERC Act 2006 S41.
Protected species in and around the site	After using the PEA desk search as a screening for records in and around the site, walked coverage should extend 500 m beyond site boundaries.
Age of data attached to the application	Field data are out of date within two years (CIEEM 2019; Natural England 2022o), but Natural England (2022o) accepts data no older than four survey seasons, and qualifies this by stating that if the development is expected to have a negative effect on GCN, then data must be more recent.
Objectives of surveys	Because of its European Protected Species status, any development that risks disturbing GCN needs to be sure of the potential impacts. Objectives will include: – presence/absence of suitable habitat on the site or within 500 m – enough to stimulate a detailed survey or seek involvement in the DL process – assessing which areas on a site are used by GCN – assessing GCN population sizes – assessing potential habitat loss and management effects on GCN – assessing potential effects of removing populations in case of partial/total area destruction. Subsequent to any development, a round of monitoring surveys would be expected, with these matching techniques pre-development. These would be as part of a mitigation plan (Natural England 2022o). For many sites, there may well be important cumulative effects to consider.

Species: Great Crested Newt

Field collection methods	Where the DL route is not taken, there are a range of methods available for GCN, usually ending in an application for a mitigation licence. 1. Environmental DNA: eDNA As GCN (young and old) shed DNA into the water, it is possible to undertake a one-off set of samples to detect their presence/recent presence by sampling water from a pond or waterbody by taking twenty 30 ml water samples from around the edge of a pond, taking care not to disturb the sediment (eDNA can be preserved in sediment, leading to a false positive). After processing at accredited laboratories, eDNA results may be obtained. eDNA is unable to provide density/populations estimates. The time window for eDNA is short (mid-April to late June) (NE 2022), although positive results may be obtained beyond the period; negative results may then not necessarily equate to absence (CIEEM 2020; Perrin 2023). If not using presence/absence (p/a) techniques, such as eDNA, then there are a range of established standard techniques that require longer periods to provide population estimates or likely absence. Sewell et al. (2013) identified five survey types: pitfall trapping, bottle trapping, netting searches, torch searches and visual searches. As GCN are European Protected Species, licences are needed for trapping or netting. With a focus on presence in or around waterbodies, the timing of all five is constrained to a few months. 2. Pitfall traps Pitfall traps involve channelling animals via up to 20 m of part-buried drift fencing into pitfall traps with grass/hay at the bottom and a 'raft' in case of flooding – unless drilled for drainage. A small mammal ladder should allow shrews to escape. With up to 20 traps, they should be checked in the early morning every 24 hours, or more often (Gent and Gibson 2003) over a period of up to 20 days. Given the level of input required, pitfalls are normally associated with mitigation projects. 3. Bottle traps Based on part inverted plastic bottles fixed by a stake, bottle traps allow ready entry by newts, but are less easy to leave. Placed around at 2 m intervals, or in deep water, overnight or for up to four hours in the early morning, bottle traps can be used to estimate population size classes. The length of time allowed before emptying decreases across the summer as temperature rises (Gent and Gibson 2003). At least three sessions are needed March to May. At 2 m intervals, catch rates may be between 2 and 28% of the pond population (Gent and Gibson 2003).

(continued)

Table 5.16 Species guidance: Great Crested Newt (*continued*)

Species: Great Crested Newt

4. Egg searches

   Newt eggs are laid individually on leaves in a waterbody from April to June. GCN eggs are larger than other newts. While confirming presence, they cannot infer relative population size. One to six visits are needed to confirm presence.

5. Netting

   These are used March to May when adults are in the pond, or mid-July to August when larvae are large; netting is a relatively inefficient method of detecting newts. Sweeping a 2 to 4 mm mesh net through the edges of a waterbody risks damage or disturbance. Natural England (2001) and Gent and Gibson (2003) use it solely to confirm presence. Fifteen minutes of sampling per 50 m of shoreline are suggested (CIEEM 2020), with a minimum of 4 to 6 visits. As adults move into deep water during the day, netting is best done in the early morning.

6. Torch survey

   This requires walking slowly around a pond in the dark March to June on warm, still evenings in non-rainy periods at temperatures >5°C with a 100,000+ candle torch, scanning every 2 to 3 m. A minimum of four visits is needed for presence/absence, and six are needed to begin to see population structures.

   Comparisons

   Langton et al. (2001) recommend a combination of torch searches, netting and egg searching to determine presence/absence, with a 90% likelihood of GCN detection. Langton et al. (2001) also recommend a mix of torch searches, netting and bottle trapping for relative abundance estimation. Note, water temperature affects detection, with more visits needed in cooler than warmer periods (Sewell et al. 2013).

7. Terrestrial searching

   In addition to the pond-based methods, Froglife (2001) and Gent and Gibson (2003) note the use of searching under refuges (e.g., rocks, logs or rubble). This extends the survey period into March to October but has low reliability (Sewell et al. 2013).

   Population estimation

   Methods 1 to 7 provide qualitative data. NE (2001) indicated a minimum of 100 days of drift fencing and pitfall traps, and a minimum of 20 days of netting and bottle trapping to begin to estimate a specific population. Sewell et al. (2013) used a known population of marked GCN to test this. Both drift and bottle methods missed known individuals. Long-term bottle trapping and statistical programmes may provide annual catch rates and population estimates. This would be outside of planning requirements.

Species: Great Crested Newt

Survey timing	Method	Jan	Feb	Mar	Apr	May	Jun	Jul	Aug	Sep	Oct	Nov	Dec
	Bottle		+	*	*	*	+	+	*(L)	+(L)			
	Egg			+	*	*	*	+					
	Torch		+	*	*	*	+	+	+(L)	+(L)			
	Net		+	*	*	*	+	+	*(L)	+(L)			
	Pitfall		+	*	*	*	+	+	+	*	+		
	Refuge			+	*	*	*	*	*	*	+		

* = most effective; + = less effective; (L) = detection of larvae. Source: Natural England (2001)

Timing of visits in the day	1. eDNA: no best time. 2. Pitfall traps: check between 06.00 and 11.00 daily. 3. Bottle traps: leave out no longer than 12 hours from March to April, 10 hours in May, 8 hours in June and 7 hours in July and August. Check 06.00 to 11.00. 4. Eggs: search in good daylight. 5. Netting: early morning. 6. Torch: in dark conditions. 7. Terrestrial: in good light conditions.
Visit frequency	1. eDNA: a single visit between mid-April and late June. 2. Pitfall traps: 60 to 100 days (NE 2001). 3. Bottle traps: where detectability is high, four visits; where low, six visits (Sewell et al. 2013). 4. Eggs: one to six, until eggs found. 5. Netting: as (3). 6. Torch: as (3). 7. Terrestrial: 60 (Hill et al. 2005).

(continued)

Table 5.16 Species guidance: Great Crested Newt (*continued*)

Species: Great Crested Newt

Survey duration	Duration varies with methods and sites.
Weather	Warm, dry, low wind speeds; avoid previous rain periods. Cold periods depress GCN water temperatures and activity (Sewell et al. 2013), and prolonged dry periods negatively influence recording.
Light	Light conditions influence detectability, behaviours and water depth use.
Distance	All areas of the site should be sampled, according to the habitat types, with habitats 500 m beyond a waterbody.
How far beyond development boundaries	500 m is the basic search radius.
Years of data	At least one year's worth of data is required. It is expected that data submitted will be adequate for determination purposes, unless an LPA has sufficient information to assess the application without this data, in line with licensing policy 4. (Natural England 2022o) This allows for developers to propose worst-case scenario compensation in certain circumstances.
Limitations	GCN surveys are affected by a wide range of issues, including light, weather, season, temperature (air and water), survey timing, time of day, preceding weather conditions, surveyor experience, number and spread of visits across the active season. There is a risk of false negative and false positive results from eDNA (Harper et al. 2019).
Suitable licensed and experienced surveyors	To work on GCN and risking disturbance requires a mitigation licence (level 2). CIEEM expects a high level of training in its GCN competency document (CIEEM 2013).
District level licensing	A site, unless it is a red risk site, may be suitable for either a DL scheme or, if the developer so chooses, determination of impacts via a mitigation licence approach. District level licensing (DLL) is an alternative approach to traditional mitigation licences (Natural England 2022o) when developing sites which could affect Great Crested Newts. It aims to increase numbers of Great Crested Newts by providing new or better habitats in targeted areas to benefit their wider population. Natural England (2022o) describes DLL as a simpler, quicker process for developers than applying for a mitigation licence. If there is a DLL scheme in the area, developers can simply apply for a DLL licence and pay for compensation ponds.

Species: Great Crested Newt

If the developer chooses not to use a DLL where one exists, LPAs need to understand the possible risks based on:

- distribution and historical records suggesting GCN may be present;
- if there is a suitable waterbody, such as a pond or ditch within 500 m of the development, even if it only holds water for some of the year;
- the development site includes refuges, such as log piles, rubble, grassland, scrub, woodland or hedgerows within 500 m of suitable aquatic habitats (static or slow-moving waterbody)

and developers should submit qualitative and quantitative information with their planning application on how their development avoids or mitigates harm to GCN. If there may be a negative effect on GCN, the LPA needs to consider the following factors to assess the site's importance:

- the number of GCN populations and their size
- the nature of the population: for example, if the site includes a breeding waterbody or is connected to other important populations;
- how important the site is to the local and national GCN population: for example, how near it is to an SSSI where GCN is a listed species;
- other GCN habitats within 500 m proximity of other ponds is critical, as ponds can reduce value by disturbance, pollution, loss of habitat connection and lack of access for management.

If the developer progresses through the DLL route, developers apply for an impact assessment and conservation payment certificate (an IACPC) from Natural England.

Natural England will:

- measure the impact of the proposed development on Great Crested Newts;
- assess the cost of compensating for the impact through new or improved ponds for Great Crested Newts;
- issue an IACPC if the development is suitable for district level licensing;
- countersign the IACPC once the developer has signed and returned it.

(*continued*)

Table 5.16 Species guidance: Great Crested Newt (*continued*)

Species: Great Crested Newt	
	Some developers may need to pay a first stage conservation payment to Natural England before they can receive the countersigned IACPC. Developers can then submit this document with their planning application. If a developer does not submit an IACPC or an alternative assessment of risk, the LPA needs to be sure that where risk is acceptable and a DLL possible (amber or green but not red) a mitigation licence should be sought if DLL is not used.
	Under the DLL, a detailed GCN assessment is not needed in an Environmental Impact Assessment, but there will need to be an entry on impacts and compensation in the Environmental Assessment (ES), matching that in the IACPC.
Mitigating properly assessed potential impacts	Mitigation is included in a DLL: • for the mitigation licence route, where ponds are to be destroyed, compensation will include: replacing the lost pond with at least two new high-quality ponds of the same area or more, on the development site; • make sure the new ponds are ready for GCN before the old pond is destroyed: this is likely to take at least one full growing season; • safeguard or replace other ponds which may be used by GCN within 500 m; the developer will need an EPS mitigation licence (Natural England 2022o) to do this.
Licensable activities	Disturbance to a European Protected Species requires a level 2 GCN species licence and a subsequent mitigation licence.
Monitoring	Under DLL, monitoring is required as part of a plan. For sites not in a DLL scheme, there should a site monitoring and management plan. These measures are likely to be needed by protected species licences. NE (2020) suggested that a site management plan could include: • aquatic vegetation management in ponds • clearance of shading tree or scrub cover around pond margins • desilting and clearance of leaf-fall • mowing, cutting or grazing of grassland • woodland and scrub management.

Species: Great Crested Newt	
	A maintenance plan should deal with: - the effects of introducing fish to the pond - pond leakage - dumping of rubbish - fires or other damage - damage to fences - tunnel silting or blockage - damage to interpretation boards. The plan should monitor newts and their habitats, including ponds, after development. It should include carrying out management works to habitats and additional survey work to check that mitigation measures are working as intended, followed by remedial work if needed.
Summary	A PEA is required, with field and desk data. Times for field visits vary with species. Reliability of desk data is uncertain. A minimum of one year of visits is required. Survey methods vary with species and objectives/development type. Surveys are affected by a range of factors including weather, time of year, time of day, surveyor experience, scale, visibility, distance and temperature, visit timing and number of visits and season. GCN can be approached via either the DLL or mitigation licence routes. The efficacy of GCN mitigation is uncertain (Nash et al. 2020; Hunter et al. 2021). Wider coverage than the site alone is needed to fully understand potential population impacts. Mitigation needs to assess habitat impacts, and if unavoidable, will require site management. Any translocation requires pre-prepared suitable habitats with long-term security and funded plans in place.

GCN = Great Crested Newt; DL = district licences; DLL = district-level licensing; IACPC = impact assessment and conservation payment certificate

6. Using the Data from Effective Protected Species and Other Surveys

Given the adversarial planning system in the UK, especially in England, it would make sense to base any argument for or against a development on fit-for-purpose data – data gathered by the correct use of standard methods. Correct is a matter of opinion, perhaps. As the NE January 2022 SA protected species guidance refers to the use of standards, both CIEEM's list of survey references of good practice guidance (CIEEM 2021) and how to approach the collection of evidence-based surveys, data collection and interpretation and proofs BS 42020 (BSI 2013), it would help if these were applied.

Both BS 42020 (BSI 2013) and CIEEM (2021a) and CIEEM's EcIA guidance (CIEEM 2019) include the need to identify and report limitations and their effects on impact assessment. That is also expected to be applied by consultants as part of CIEEM's ethics policy, signed up to when joining CIEEM.

It was noted earlier that limitations should be one of the longest and best documented and argued sections of a planning submission. Instead (Reed 2021) it is one of the shortest and usually has no recognition of the precepts in BS 42020 (BSI 2013) that the NE SA requires. Do LPAs themselves recognise the need for limitations to be clearly identified? If they do, that will help in weeding out the good from the poor. Some do: Chichester District Council (2018) states that:

> A survey will need to:
> - Record any uncertainties and limitations that might affect survey results. (Chichester District Council 2018)

For applications that fail that test, they risked being kicked back by the Council.

Camden Borough Council (2018) goes further, setting out what it sees as limiting factors, and how these work in practice: https://www.camden.gov.uk/documents/20142/4823269/Biodiversity+CPG+March+2018.pdf/daf83dad-d68d-6964-99b4-aef65d639304?t=1562591072964

For others, such as Dorset County Council (2023), the guidance is much more implicit. Rather than choosing to follow Camden or Winchester, Dorset instead refers applicants requiring support to the NE SA guidance. As noted above, that might not be a wise idea: https://www.dorsetcouncil.gov.uk/documents/35024/283704/Ecological+Survey+Season+Guide.pdf/4995a402-28e3-de05-909e-785181280b65.

For Essex County Council (2015), there is no mention of limitations in its VC and stated requirements. Instead, it refers back to the NE SA guidance for applicant support: https://www.placeservices.co.uk/media/108412/biodiversitychecklist-15essex2018.pdf

For some, such as North Lincs District Council (2020), the delivery of data for net gain is done without reference to limitations at all: https://localplan.northlincs.gov.uk/stages/3/9-environment:

1. minimise and mitigate against impacts on biodiversity and geodiversity where adverse effects are unavoidable; and
2. deliver a net gain in biodiversity and/or geodiversity.

In a review of how 352 LPAs approached No Net Loss (NNL) and Net Gain (NG) between May 2019 and December 2020, Robertson (2021) came up with similar points:

- There was little consistency in how LPAs approached NNL or NG. Only 32% of LPAs considered NNL/NG mandatory.
- 34% of LPAs used some sort of metric; 62% did not. Most of those that did – but not all – used DEFRA BNG version 2. It, and other versions of DEFRA's BNG are not comparable, and certainly not additive (BNG 3.1 (Panks et al. 2022)), so that LPAs risk producing outputs that are chalk and cheese.
- To assess NNL/NG, 80% of LPAs used professional judgement, applicant submissions or did not use evidence at all.
- Where conditions were attached to permissions there was little or no enforcement/ follow up, so that NNL/NG were not effectively assessed.

With so much variation in the official LPA line, it appears that getting a hard and fast set of inter-council comparative data to begin to approach something that might produce a national level figure for positive gain/NNL is going to be tricky in practice.

The need for securing measurable nets gains is being driven in part by the NPPF. The NPPF (HMG 2021) paragraph 179 states:

> promote the conservation, restoration and enhancement of priority habitats, ecological networks and the protection and recovery of priority species; and identify and pursue opportunities for securing measurable net gains for biodiversity and development whose primary objective is to conserve or enhance biodiversity should be supported; while opportunities to incorporate biodiversity improvements in and around developments should be encouraged, especially where this can secure measurable net gains for biodiversity.

The use of the term measurable is the first difficulty in paragraph 179. As seen above, getting good survey data with a high degree of reliability is difficult. For protected species, the detailed standards make reference to the problems of estimating reliable datasets. This means that some assessment of the error term in estimates might be expected. Error terms or ranges are singularly missing from almost all population estimates accompanying planning submissions. Normally, limitations are either missing or unsubstantiated. Studies that looked at estimates of counts and comparisons of survey results indicated that the level of error is high (e.g., Cherrill 2013, 2016; zu Ermgassen et al. 2021) and the error is quite likely to be greater than the gain claimed by some gain estimates. BNG 3.1 (Panks et al. 2022) paragraph 1.9 also comes to much the same conclusion: 'The quality and reliability of outputs will depend on the quality of the inputs.'

The NPPF (HMG 2021) ill-advisedly melds the status of priority and protected species with the concept of biodiversity gain. As BNG (Panks et al. 2022) itself notes (see below), the BNG Tool is a habitat estimator and does not cover protected or priority species.

If a measurable net gain is to be achieved by a development involving protected species, then it is critical that, as well as there being good population and area estimates, the mitigation options being advocated are suitable. If they are not (Hunter et al. 2021), or cannot be measured, or, as Robertson (2021) showed, are not measured, then net gain is implicitly an illusion. Rampling et al. (2023) indicated that the level of error in BNG calculations (>10%), problems with habitat recognition and lack of LPA capacity to assess what is being claimed all contribute to problems with assessing credibility.

The NE SA was clear about the need for positive mitigation, and each of the guidance documents contained suggested options. Assuming that there is an estimate for the extent of mitigation that is required, and that can be relied upon – and that may be contentious (Hunter et al. 2021; Drayson and Thompson 2015) – then the scale and reliability of the mitigation options will be critical to delivering a net gain.

Most planning applications show a range of essentially generic actions that are to be used to enhance or improve a site for biodiversity. Bird boxes and bat boxes (Lintott and Mathews 2018) are a common option in many lists. The key question to ask is: do these work? Which work and to what scale?

As noted earlier, Hunter et al. (2021) reviewed how ecological mitigation and compensation measures (EMC) are used by large-scale development projects to mitigate their impacts. Looking at 50 big housing developments, Hunter et al. (2021) found 446 measures were put forward, covering 65 actual EMC measures – on average, nine measures per development application. Given the scale and importance of the 65 EMCs, Hunter et al. (2021) asked whether or not there was any basis for the measures. Did they work? Was there any empirical data behind them to back them up. In essence: were they fact or fiction, or just undocumented acts of faith?

As they started to unpick the datasets, the first thing they found was that where there had been comprehensive reviews of measures undertaken, few of the measures had been found to be evidence-based and even fewer had post-development monitoring. These gaps were quite critical. For example, Lewis et al. (2016) found no published literature supporting Great Crested Newt mitigation actions. For another EPS species group – bats – Stone et al. (2013) found significant reductions in bat abundance in

300 derogation licences, contrary to expectations. Lintott and Mathews (2018) found just 52% of lofts built as compensation activities actually contained bats, and only 31% of bat boxes that were used as the main mitigation option were occupied. Neither is an overwhelming vindication of the EMC approach. Nor would biodiversity gain be realistically achieved.

Looking a little more deeply, Hunter et al. (2021) asked what evidence there was to back up the mitigation options being advanced in support of the actions meant to deliver 'measurable' net gain: was there any? Did the actions taken actually get followed up to confirm they provided a measurable return? Did they work?

Probably the best sort of evidence of EMC, if it is not possible to choose sites randomly, is the before–after control–impact approach (BACI), where sites with impacts are compared with similar sites with natural processes. Only two cases were available in all EMC references. Of the 52 unique measures for EMC, 30 (58%) had no evidence or were not assessed. Delving further into the measures cited, more than half were citations of other citations, which themselves did not have a firm evidence base. As Hunter et al. (2021) observed, the lack of evidence, and what they termed circular referencing – the citing of non-evidenced citations – means that EMC measures were propagated without any proof of validity. For example, 'destructive search' where vegetation and soil are stripped to identify remaining animals on a site was widely cited and recommended in 18 reports. Yet, it lacked any empirical basis. Similarly, Nash et al. (2020) found no confirmatory evidence for reptile translocation as a basis for mitigating development impacts.

Hunter et al. (2021) concluded that, with <10% of the evidence used to support guidance recommendation coming from empirical evidence and ecological decisions being made on unsupported 'standard practice' and unsupported professional judgement, for LPA and thus national biodiversity targets to be met, avoidance was normally the only credible option. Failing to do so would be to implement measures that had not been shown to be effective. That is not a recipe for delivering NNL, biodiversity gain, or BNG.

6.1 Biodiversity Net Gain and species

If the very material that should be used to assess the mitigation cases for potential impacts is of limited reliability, where does this place the use of BNG in future policy development? It is clear that the version of BNG going to be used for the next 20 plus years (BNG 4 (Natural England 2023)) is not suitable for protected species, nor is there any obvious basis for presuming that a set of hoped-for linked species will necessarily be delivered along with a habitat type. If the mitigation methods are largely baseless – in more ways than one – then BNG is playing on what sportspeople call a sticky wicket – anything might happen.

The BNG project has gone through a series of iterations, and the accompanying guidance has altered along with it. The most recent BNG 4.0 user guide (Natural England 2023) sets out the 'rules' that must be followed. These are followed by equally succinct 'principles', rather than the other way round. It is a terse document, and one where protected species are not mentioned at all.

For clarity, and to understand what BNG hopes to achieve, it helps to look at the user guidance for version 3.1, as well as for version 4.0.

The clearly stated objective of BNG 3.1 (Panks et al 2022), a rather more expansive, helpfully explanatory version, now replaced by version 4.0 in March 2023, was to leave biodiversity better than when a development was planned – where better meant that biodiversity would be at least 10% higher than the baseline assessment. This was to be done by scoring the habitat-based 'biodiversity units' before and after the development. These can be inside the development, by 'improving' habitats such as grassland, or delivered in external offsets. Such changes will not be instantaneous, but may occur over time as habitats mature and increase in ecological value. It omits assessment of numbers and identity of protected species on the site.

Like the SA, and the studies by zu Ermgassen et al. (2021, 2022a, 2022b) and Rampling et al. (2023), the viability of the BNG process depends on data quality: the ability of field staff to produce credible data. Both BNG 3.1 and the user guide for version 4.0 stated clearly: 'The quality and reliability of outputs will depend on the quality of the inputs.'

Without reliability, the calculation of BNG scores and 10% improvement is at risk (zu Ermgassen et al. 2021, 2022a, 2022b). For major projects, such as the HS2 rail line, the very reliability of the data used to make its major claims of NNL and BNG are at issue (The Wildlife Trusts 2023). This is fundamental, even if attempting not to use protected species data in their claims.

Understanding what BNG does or does not bring with it is critical. As BNG 3.1 paragraph 1.10 (Panks et al. 2022) stated:

> It does not include species explicitly. Instead, it uses habitats as a proxy. The metric does not change existing levels of species protection and does not replace regulatory processes for species protection.

In BNG 4, that reference to protected species had been removed, but there was still an element of caution in Table 2.1: 'Biodiversity units are a proxy to describe biodiversity',

where in 1.5.5 it stated:

> the outputs of this metric ... provide a proxy for the relative biodiversity worth of a site pre-and- post intervention ... The metric should be used alongside ecological expertise as part of the evidence that informs plans and decisions.

Without formal mention of protected species, and a sideways reference to rarity and species richness under the use of the term 'distinctiveness' (BNG 4: Table 2.1), achieving protected species gain and positive safeguarding through BNG may well be a happy accident, rather than achieved by design.

As a result, LPAs should not confuse protected species outcomes with BNG outcomes. They are not the same, and should not be treated as such. As noted above, as the capacity and the ability of LPAs to recognise protected species is limited, and NE guidance is inadequate, then there is a clear risk that threatened or protected species could be lost while a site achieves an apparent habitat net gain.

BNG 3.1 (Panks et al 2022) stated some of the limitations that a user should know, and appreciate, before embarking on using the BNG tool. These included:

> Principles and rules for using the metric
>
> 2.18. The metric is a tool that can be used to help inform plans and decisions. It is important, however, to be aware of its limitations and to conduct assessments with regard to a set of principles and rules.
>
> Limitations
>
> 2.19. The metric uses habitats as a proxy for biodiversity ... it is a simplification of complex ecological processes which are not readily captured ... the outputs of biodiversity unit calculations are not scientifically precise or absolute values. Therefore, the generated biodiversity unit scores are a proxy for the relative biodiversity worth of a habitat or site.

That statement is simplistic: if species data are mainly absent, then a key component of that relative worth is also missing.

> 2.20. The metric and its outputs should therefore be interpreted, alongside ecological expertise and common sense, as an element of the evidence that informs plans and decisions. The metric is not a total solution to biodiversity decisions ... it does not tell you the appropriate composition of plant species to use or which micro-habitats might benefit locally important species.

The caveated role as a proxy, and its limited capacity in biodiversity decision-making (the absence of species details), need to be borne in mind at all stages of BNG evaluation. It is all too easy for a decision-maker to take the number of habitat units generated by the BNG tool as absolute, not aware of the simplifications or problems that hide behind a figure.

The principles set out for use of BNG 3.1 include:

> Principle 3: The metric's biodiversity units are only a proxy for biodiversity and should be treated as relative values. While it is underpinned by ecological evidence the units generated by the metric are only a proxy for biodiversity and, to be of practical use, it has been kept deliberately simple. The numerical values generated by the metric represent relative, not absolute, values.

The risks that come from using a proxy for unnamed and unquantified species details cannot be overstated. The species flag is further fluttered in Principle 4, but not addressed in the metric:

> Principle 4: The metric focuses on typical habitats and widespread species; important or protected habitats and features should be given broader consideration.
>
> - Protected and locally important species needs are not considered through the metric, they should be addressed through existing policy and legislation.

This raises the potentially rhetorical point, that all of the recent – 2022 – NE protected species guidance states the duty of the LPA to achieve a net gain in biodiversity through good design. What might be appropriate, and how this fits with BNG Principle 5, is uncertain:

> Principle 5: The metric design aims to encourage enhancement, not transformation, of the natural environment. Proper consideration should be given to the habitats being lost in favour of higher-scoring habitats, and whether the retention of less distinctive but well-established habitats may sometimes be a better option for local biodiversity.
> - Habitat created to compensate for loss of natural or semi-natural habitat should be of the same broad habitat type (e.g. new woodland to replace lost woodland) unless there is a good ecological reason to do otherwise (e.g. to restore a heathland habitat that was converted to woodland for timber in the past).
> - Although the metric does not explicitly consider the biodiversity value provided by individual species, consideration should be given to locally relevant species interests when creating or enhancing habitats.

Even if protected species are not formally addressed by the BNG tool, they may well be affected by its users, inadvertently at best. The principles for guidance version 3.1 were condensed in the user guide for version 4.0 (Natural England 2023) (Table 6.1).

6.2 Testing BNG

A tool such as the BNG does not arrive overnight. Rather, it goes through a long process of testing and evaluation. This helps to show its reliability, plug gaps and increase its general credibility.

Several studies have looked at the data collected by versions of BNG and noted a series of issues that need to be borne in mind.

The first question, harking back to BNG 3.1 paragraph 1.9, is that of data quality: poor data risk upending the basis of the calculations. The BNG relies on field workers correctly identifying the habitat type and its condition. 'Improving' habitats and their condition is at the heart of the 10% claim. Zu Ermgassen et al. (2021) looked at trials of BNG 2 at six early-adopting LPAs. Starting with the consideration of paragraph 1.9, that quality and reliability of the data determine BNG, they trialled the ability of seven expert grassland ecologists to recognise the 'correct' grassland type and condition score, both pretty basic requirements for baseline assessments and for subsequent habitat scores and BNG changes. The experts agreed with both type and condition 31% of the time, with type 42% of the time and condition alone 62% of the time. It might help to put those data in context: tossing a coin would produce a 50% chance of agreement, so to be 30–40% likely to agree is not confidence-building.

If experts could not agree consistently, then given the earlier noted limits of LPA staff (Robertson 2021; Snell and Oxford 2022) there is little chance that LPAs would have the capacity to reliably scrutinise type and condition assessments, the very building

Table 6.1 Biodiversity metric principles

Principle number	Principle detail
Principle 1	This metric does not change existing biodiversity protections. statutory obligations, or policy requirements.
	The use of this metric does not override the ecological mitigation hierarchy and other requirements (such as consenting or licensing processes. for example woodlands).
Principle 2	This metric should be used in accordance with established good practice guidance and professional codes.
Principle 3	This metric is not a complex or comprehensive ecological model and is not a substitute for expert ecological advice.
Principle 4	Biodiverstty units are a proxy for biodiversity and should be treated as relative values.
Principle 5	This metric is designed to inform decisions in conjunction with locally relevant evidence, expert input, or guidance.
Principle 6	Habitat interventions need to be realistic and deliverable wtthin a relevant project timefraine.
Principle 7	Created and enhanced habitats should seek., where practical and reasonable, to be local to any impact and deliver strategically important outcomes for nature conservation.
Principle 8	The metric does not enforce a minimum habitat size ratio for compensation of losses. However, proposals should aim to: • maintain habitat extent (supporting more, bigger, better and more joined up ecological networks} and • ensure that proposed or retained habitat parcels are of sufficient size for ecological function

Source: BNG version 4 user guide (2023).

blocks of the BNG process. Marsh (2022) calculated that for his Leeds LPA, in order to properly implement BNG as envisaged by the government would cost an additional £320,000 annually. That appears unlikely to be resourced.

If experts cannot agree, and LPAs lack the capacity, capability and money to assess sites, then how BNG can be monitored – for it falls on LPAs to do this (Robertson 2021; zu Ermgassen et al. 2021, 2022a, 2022b; zu Ermgassen 2021; Marsh 2022) – is also a problem.

Even if the LPA does monitor BNG delivery, it is argued that enforcement would not be taken unless the violation was a serious harm to a local public amenity (zu Ermgassen 2022b; Rampling et al. 2023) and that this would be unlikely to be triggered (House of Commons Library 2019). Effectively, remedies to deliver failed, but promised, biodiversity gains would be unenforceable (zu Ermgassen et al. (2022b). Drayson and Thompson (2015) reviewed outcomes of mitigation actions linked to approved developments. Few of them were implemented and fewer still were followed by LPAs. In addition, where monitoring has been self-reporting, rather than third party led, standards have been poor. Robust evidence of the delivery of habitat units is required, but is lacking (zu Ermgassen et al. 2022b). Rampling et al. (2023) noted that a fifth of all metrics data they looked at for version 3.1 had errors, and almost all appeared to have been unnoticed by LPAs. These would have been picked up by a well-resourced LPA, something that Robertson (2021) noted was almost absent.

It looks like BNG is a good idea, but is based on partial data and partial reliability. It cannot be relied upon to deliver its aspirations, and from a protected species standpoint it is largely incidental.

6.3 Delivering protected species into the future

If there really is an intention to properly include protected species in biodiversity planning and development in the future, then what we have now is not going to deliver: for species, for planners, for biodiversity or for government, no matter what it says it means: now or in the next 25 years (Gove 2018).

To help change this, the first step will be to make available a clear guide to planners that allows them to understand the sorts of survey data they need and how to differentiate between good and bad planning submissions for individual protected species. Hopefully, this guide goes some of the way to achieving that goal.

The current set of NE SA (Natural England 2022a–o) is not suitable for the purpose and roles that it claims to represent. For that reason, the 2022 SA needs to be revised so that it meets the needs of LPAs, and also clarifies the real-time expectations of consultants too.

ALGE needs to improve its documents and to update the guidance that it and its partners have produced, so that these can help fill the capability gap in its overburdened staff.

The LPAs are staffed by hard working, but totally overburdened planners and ecologists. As all of the ALGE surveys show, they are unable to deliver their basic mandatory functions with their existing staff levels. Adding on more demands to fewer staff is unreasonable. Marsh's (2022) estimated additional (approximate) £300,000 for his own LPA to begin to properly deliver BNG as set out by government cannot be too far off the mark for many LPAs and far less than counties or LPAs will need.

Being able to provide standard, properly collected and evaluated data to LPAs would make the life of LPA planners and their ecologists more tolerable. For them in turn, being able to properly undertake their role would allow the production of data that could be meaningfully comparable for use at regional and national levels.

As zu Ermgassen et al. (2021), Rampling et al. (2023) and others (Cherrill 2013, 2016; Hearn et al. 2011) have shown, for any effective programme of biodiversity delivery or policy assessment, the buck stops with data reliability. If consultants and LPAs both work with good-quality data, collected well, with clearly stated limitations, then planning decisions can be made that have meaning.

Continuing to work with poorly collected data, which do not fit agreed standards (such as BS 42020), does nothing for the profession of ecology, for planners and for their ultimate client: sustainable biodiversity.

Failing to use a guide such as this to help improve performance leaves a door wide open for the continued drain and piecemeal loss of biodiversity, and confirming England and the UK as one of the most nature-deprived areas in Europe and beyond. Nobody wants that, but it is our prize unless we actively choose to do otherwise.

Appendix

Key elements for each species/species group in both the pre-2022 NE SA and PBP SGN species guidance, with comments on gaps/coverage compared with the standard reference for that species.

Species/group and date of pre-2022 publication	Timing			Field methods			Impacts		Mitigation		Post-development monitoring			References
	Introduction to SA initial Table (1) times	SA species detailed text: times	SGN times	SA field methods	Standard reference	SGN field methods	SA impact factors list	SGN impact lists	SA mitigation method	SGN mitigation methods	SA	SGN	SA	
Badger 28 3 2015	Any time of year setts. February to April and October to November for bait marking	Any time of year setts; best early spring or late autumn	Sett survey best February to April and September to November. Bait marking February to April	Setts: occupancy bait marking 4 weeks. Signs: sett; active soil; bedding; footprints; paths; faeces; hairs, scratching posts; digging	Sett check October to March in 1 km search area. Signs: sett; faeces; setts; nest material; hair; footprints	Sett occupancy bait marking 4 weeks. Signs: faeces; setts; paths, footprints. Camera traps	Sett damage Sett loss Foraging area loss Disturbance	No formal list; move straight to mitigation	Use mitigation hierarchy avoidance. Retain/make new foraging areas. Maintain connectivity. Short-term exclusion. Artificial setts	Use mitigation hierarchy. Sett closure. Restore routes. Underpasses. Artificial setts	Not identified	Yearly or twice yearly, for several years if mandated	None given	4 references
Comments	No month details; no distances. Limitations not addressed	No month details; no distances. Limitations not addressed	No distances. Limitations not addressed	No distances; no frequency for checks – bait marking times unstated in particular. Limitations not mentioned		No distances; frequency for bait or cameras unstated. No limitations mentioned		No impact list	Durations not detailed	Some dates mentioned for closure only	No post-mitigation monitoring mentioned	Need for budgeting for monitoring noted	None	Not clear which is most important

Species/group and date of pre-2022 publication	Timing			Field methods			Impacts		Mitigation		Post-development monitoring		References	
	Introduction to SA Initial Table (1) times	SA species detailed text: times	SGN times	SA field methods	Standard reference	SGN field methods	SA impact factors list	SGN impact lists	SA mitigation method	SGN mitigation methods	SA	SGN	SA	SGN
Bats 28 2 2020	Preliminary roosts: anytime. Hibernation roost: November to mid-March. Summer roost: May to August. Swarm August to October.	Preliminary roosts: anytime. Hibernation roost: December to February. Spring; Summer: March to May. Summer roost: May to September. Swarm: August to October	Preliminary roosts: anytime. Hibernation roost December to February. Summer roost: May to August/October. Swarm: August to September/October	Use records as well as field assessment to decide on surveys. Methods include: visual; bat detector; net; harp trap; radio tracking	BCT details methods.	No reference to desk data or PEA walkover. Methods almost all visual or bat detectors	Summary of how to assess roost importance and impacts listed: short- and long-term impacts identified. Detailed effects on roost types listed	No formal impact section	Avoidance, then other parts of mitigation hierarchy. Methods listed	Mitigation hierarchy. List 3 avoidance options then detailed mitigation according to locations (trees, buildings, turbines and roads). Enhancement included	Monitoring plan should be put in place. Turbine reference given	Post-implementation monitoring. Identified as important and could be up to 10 years	BCT and bat turbine. References cited	Exhaustive list provided. Divided into topics
Comments	Hibernation roost times differ from introduction to SA	Some dates differ from BCT guide	Dates vary from BCT and where in UK	Details cursory, and not in line with BCT; licensing issues omitted. eDNA omitted as method	Both SA and SGN refer to BCT, but actual details differ in each	Basic details of timing, frequency and survey poor. Limitations missing	Lists more detailed than methods	Details only in mitigation hierarchy	Expanded details in BCT	Lists adequate, no time or numerical details	Too basic to use without referring to references.	Too basic to use; references needed	BCT a complex document. Turbine details complex and hard to understand	No guidance as to which reference is best to use. Confusing for LPA user

Species/ group	Timing			Field methods			Impacts		Mitigation		Post-development monitoring		References	
	Introduction to SA initial Table (1) species	SA species detailed text: times	SGN times	SA field methods	Standard reference	SGN field methods	SA impact factors list	SGN impact lists	SA mitigation method	SGN mitigation methods	SA	SGN	SA	SGN
Birds 28 3 2015	Breed: March to August Winter: October to March. Passage: March to May and August to November	Breed: March to August	Breed: April to June. Winter: November to February. Pass: October to March. Times vary between text sections	Scope need: get records and do walkover. Limited list of habitats given	Detailed generic and species-specific methods available	No desk surveys. Visit frequency related to site quality. Weather limits noted. Details for limited species range or habitat type	Damage/ remove breeding sites; disturbance of vegetation/ habitat change; building/ rockface demolition; recreation, housing or wind-farms	No impact list	Avoidance first option – list types. Mitigation and compensation noted	Detailed lists for: buildings, trees. Generic for habitats. Generic list of habitat enhancement	Need noted	Use similar methods to pre-development for up to 10 years		List provided
Comments		No details of other survey types or times		Coastal habitats missing; survey methods omitted		Times at conflict with standards. Details too short for use by LPA	Focus on breeding habitats and omit other times	Long mitigation section alludes to effects/ impacts	No details, ineffective	Habitat details too general for LPA use. Enhancement generic list	No details provided. Unsuited for LPA use		None provided; indirectly by web link to birds and turbines	Long list; priorities not clear. Wetland reference cited in text is missing

Species/group	Introduction to SA initial Table (1) species	Timing SA species detailed text: times	SGN times	Field methods SA field methods	Standard reference	SGN field methods	Impacts SA impact factors list	SGN impact lists	Mitigation SA mitigation method	SGN mitigation methods	Post-development monitoring SA	SGN	References SA	
Hazel dormouse 29 7 2015	Survey April to November	April to November	Habitat all year. Nest tubes April to November. Nest search November to March; nuts September to October	Nest tubes, nest boxes and nuts + wider area survey. For tubes, at least 50 tubes April to November. Check tubes at least every 2 months. Scoring system explained	Detailed methods available, along with scoring counts	No desk study, just preliminary visit to check for signs. Nest tube only method. Visit every 1 to 2 months	3 types given: Short-term: disturbance, increased predator risk; foraging habitat loss. Long-term: hedge loss; fragmentation; habitat loss. Post-development: disturbance and predation	No impact list or discussion	Avoidance first option – list types. Mitigation and compensation noted, include translocation or habitat management	Avoidance hierarchy. Possibly remove and translocate. Compensate with new habitat. Wide habitat enhancement	Need noted	Use similar methods to pre-development for up to 10 years	SA	List provided
Comments				Clear discussion on factors affecting survey results		More methods than set out here. Need guidance on limits and reliability	Impacts inferred		Detailed calculations for habitat clearance protocols	Habitat details too general for LPA use. Enhancement generic list with no scalar minima	No details provided. Unsuited for LPA use	Need to account for year-to-year population fluctuations	None	List provided, but no priorities/ranking

Species/group	Introduction to SA initial Table (1) species	Timing SA species detailed text: times	Timing SGN times	Field methods SA field methods	Field methods Standard reference	Field methods SGN field methods	Impacts SA impact factors list	Impacts SGN impact lists	Mitigation SA mitigation method	Mitigation SGN mitigation methods	Post-development monitoring SA	Post-development monitoring SGN	References SA	References SGN
Great Crested Newts 12 11 2020	Mid-May to mid-June. Mid-April to end June. eDNA	No dates	Recommends 4 to 6 visits in good conditions mid-March to mid-June. eDNA. Or pitfall traps for 2+ months	If not DLL, then search ponds/refuges. Survey types: Presence/absence surveys; eDNA; pond/refuge surveys within 500 m. Data <2–4 years old	DLL; eDNA; as in SA notes	Habitat potential survey visit. Search (in good conditions). 4–6 visits: dipnet, torch search, spawn/egg; bottle traps; refuge checks eDNA	Proximity: high if <50 m from pond and habitat. High to medium if ponds/habitat 50–250 m from pond and habitat. Medium: partial or temporary change to habitat. Low: temporary habitat disturbance >250 m from breeding ponds.	No formal list	If ponds destroyed: replace with 2 ponds on site. New ponds at least one year old before destroying old pond; safeguard other ponds <500 m away, with a mitigation licence	Mitigation hierarchy. Avoidance discussed and transference. Habitat enhancement includes management, ponds and protection	None for low importance, up to 10 years for high importance populations (see NE 2001 guidelines)	Spring surveys for presence, breeding and pond condition	Web links provided are invalid	A long list of references, subdivided into surveys, mitigation and monitoring
Comments	Dates differ for eDNA	Text focuses on local licensing. Limitations missing	No desk study	Timing and survey details missing, as are limitations		Distances missing; limitations of techniques not noted			Exclusion cited for <3 years if needed		No distances or times cited		No guidance on which to follow; important as they differ in detail	

Species/group	Introduction to SA initial Table (1) species	Timing – SA species detailed text: times	Timing – SGN times	Field methods – SA field methods	Field methods – Standard reference	Field methods – SGN field methods	Impacts – SA impact factors list	Impacts – SGN impact lists	Mitigation – SA mitigation method	Mitigation – SGN mitigation methods	Post-development monitoring – SA	Post-development monitoring – SGN	References – SA	References – SGN
Inverte-brates 10 8 2015	Survey April to September	May to early September	5 visits April to October	Refers to CIEEM's broken links. Refers to NBN as data source	Detailed methods available, along with ways of ranking site value	Scoping visit by specialist. No desk survey. Follow on surveys and details depend on species groups. Survey times depend on possible species. Visit in warm and dry weather	Need to assess status of species assemblage	No impact list or discussion	Avoidance by redesigning the first option; or minimise footprint. Phase work and restoration to provide habitat continuity; maintain habitats in area. Individual species have specific needs	Avoidance hierarchy. Possibly remove and translocate. Compensate with new habitat. Wide habitat enhancement	Not noted	Use similar methods to pre-development for up to 10 years	Web links to some sites (CIEEM) broken	SGN
Comments	Web Link to CIEEM broken			NBN cannot be used as a general data source for commercial consultancy. Limited value- as is SGN – for LPA use	Need individual methods for groups	No desk survey mentioned. More methods than set out here. Need guidance on limits and reliability. Focus on site quality not presence	Detailed calculations for habitat clearance protocols	Impacts inferred	Habitat details too general for LPA use. No guidance on mitigation		No details provided. Unsuited for LPA use	No mentioned. Not LPA suited	None	List provided, but no priorities/ranking

Species/group	Introduction to SA initial Table (1) species	Timing — SA species detailed text: times	Timing — SGN times	Field methods — SA field methods	Field methods — Standard reference	Field methods — SGN field methods	Impacts — SA impact factors list	Impacts — SGN impact lists	Mitigation — SA mitigation method	Mitigation — SGN mitigation methods	Post-development monitoring — SA	Post-development monitoring — SGN	References — SA	References — SGN
Natterjack Toad 12 5 2015	April to May aquatic; July to September terrestrial	April to September	None	Torch April to September; refuge: between spring to autumn; Spawn: April to early June >once a week. Calls: April to July. Count after wet periods. Data ≤2 years old	P/A: calling early April to early May. Spawn: early April to mid-June. Refugia: spring to autumn. Night lamp searching summer to autumn	None	Loss of habitats. Change in habitat management. Habitat fragmentation/isolation. Water changes. Pond shading	None	Mitigation hierarchy. Avoid sites; time work for non-pond use times. Preclude disease transmission risk. Increase connectivity. Moving to new ponds last resort	None	At least 5 years – need adults Males to return after 2–3 years. Adult females after 3–4 years. Third generation spawn	None	No formal references	None
Comments		Limitations missing	Covered in generic terms by amphibian entry	Timing and survey details missing, as are limitations. Frequency details missing for non-spawn methods		Covered in generic terms by amphibian entry		Covered in generic terms by amphibian entry		Covered in generic terms by amphibian entry		Covered in generic terms by amphibian entry		

Species/group	Introduction to SA initial Table (1) species	Timing — SA species detailed text: times	Timing — SGN times	Field methods — SA field methods	Field methods — Standard reference	Field methods — SGN field methods	Impacts — SA impact factors list	Impacts — SGN impact lists	Mitigation — SA mitigation method	Mitigation — SGN mitigation methods	Post-development monitoring — SA	Post-development monitoring — SGN	References — SA	References — SGN
Otter 5 4 2019	Any time	Any time: best is spring. Desk survey data used. Several field surveys across the year. Depends on how likely work will affect Otter; size of development	Field sign October to March. Habitat suitability all year	1. Spraints. 2. Tracks. 3. Food remains. 4. Slides. 5. Holts. 6. Couches	Preliminary survey; camera traps; road kills; dens; tracks; spraints (best May to September); couches	No desk surveys ≤250 m beyond site boundary. Signs: 1. Spraints. 2. tracks. 3. Food slides. 4. slides. 5. Holts. 6. Couches	Habitat loss. Habitat isolation/fragmentation. Loss of holts/couches. Disturbance: feeding/resting places. Changes to routes exposing to traffic. Water quality change	None	Mitigation hierarchy. Design in habitat retention. Avoid road crossings; fence out if needed. Compensate by improved habitat; better connectivity; Holt provision	Mitigation hierarchy. 1. Avoid. 2. Crossing culverts 3. Fencing to exude from sites. 4. Habitat enhancement	Not mentioned	Install checks to see if compensation works (e.g., camera traps for culverts/fencing)	None	5 but unprioritised
Comments	No details on distance or frequency			Limitations and frequency uncertain	Distances <50 m from site boundary	Limitations and uncertainty unknown	Scalars unknown		No spatial or numeric details		Absent, even though comments on routes/disturbance	Frequency, duration and distance unknown	Circular web link	

Species/group	Timing			Field methods			Impacts		Mitigation		Post-development monitoring		References
	Introduction to SA initial Table (1) species	SA species detailed text: times	SGN times	SA field methods	Standard reference	SGN field methods	SA impact factors list	SGN impact lists	SA mitigation method	SGN mitigation methods	SA	SGN	SA
Reptiles 28 3 2015	April to mid-October	April, May and September. Avoid July to August, November to February	April to June and September	Survey if PEA shows records/habitats suitable. Objective to establish population size. 1. Basking. 2. Sheets/refuges. 3. Tiles/felt	1. visual basking transect 2. Refuges. Minimum 7 visits. Rare species >11 visits required 3. March–April for hibernacula	1. Basking survey 2. Refuge checks. ≥7 visits or 11 if Sand Lizard/Smooth Snake	Loss of habitat links. Separating summer and hibernation sites. Reducing habitat quality. Fire risk. Increased litter		Mitigation hierarchy. Avoid by layout. Change work time. As last resort translocate. Compensate by habitat change / extent. Catching may take up to 3 years	Capture and transfer or exclusion. Capture 1 month or more; worst 3 years. Habitat management	Not mentioned	Surveys in spring and autumn. Check habitats for retained suitability	None	Broken into: surveys and monitoring. Mitigation etc. Management
Comments		Not same as Table 1. No frequency or spatial details	PEA but no mention of desk survey	No details on numbers, weather, limitations		Visit details: when, where/distance not set out. Limitations missing	Scalars and timescales not mentioned	None		No mention of avoidance or habitat management		No duration or scale mentioned	

Species/group	Timing			Field methods			Impacts		Mitigation		Post-development monitoring		References	
	Introduction to SA initial Table (1) species	SA species detailed text: times	SGN times	SA field methods	Standard reference	SGN field methods	SA impact factors list	SGN impact lists	SA mitigation method	SGN mitigation methods	SA	SGN	SA	SGN
Water Voles 28 3 2005	Mid-April to September	April to October	March to October; dates dependent on position in Britain	Survey if records or habitat suggest viable. Signs: droppings; latrines; feed site; burrows; footprints; runs. Look ≥2 m from water	2 visits: mid-April to June and July-end September; 2-month gap. Signs: droppings; latrines; feed site; burrows; footprints; runs Extend coverage beyond site: 50 m (small development) -500 m (big development)	Search for signs: droppings; latrines; feeding; burrows; footprints; runs. Avoid high rainfall/floods. 2 visits: mid-April to June and July to September; 2 months apart	Habitat disturbance/destruction. Disturb/destroy shelter. Change water quality.	None	Mitigation hierarchy: avoid works in vole areas. Retain connectivity. Limit habitat damage. Compensate by improving habitat, banks and water quality. Control mink	Mitigation hierarchy: avoid area; retain habitat connectivity; retain in situ habitat; buffer zone to avoid impacts. Compensate by improving water/habitat quality; enhance habitat management	No mention	Monitor mitigation suite for ≤5 years. Use techniques comparable to pre-development	None	References for: Surveys, Mitigation SGN
Comments	Desk study to be used for records		No reference to desk study	No comments on frequency/timing or distance up/down stream		Notes droppings are only reliable proof. No search distance/limitations			No buffer distance given					No priority on references

Species/group	Timing - Introduction to SA initial Table (1) species	Timing - SA species detailed text: times	Timing - SGN times	Field methods - SA field methods	Field methods - Standard reference	Field methods - SGN field methods	Impacts - SA impact factors list	Impacts - SGN impact lists	Mitigation - SA mitigation method	Mitigation - SGN mitigation methods	Post-development monitoring - SA	Post-development monitoring - SGN	References - SA	References - SGN
White Clawed Crayfish 9 10 2014	July to September	Mid-July to mid-September to avoid breeding season disturbance	Not covered	Hand netting Hand search in clear water. Night torch Licensed trapping	Trapping. Night torch. Manual search. Electro-fishing.	Not covered. Only comment in invertebrate guidance: 'appropriate surveys should be considered'	None mentioned	Not covered	Mitigation hierarchy: reduce bank disturbance. Reduce sediment release. Reduce work area. Add correct vegetation. Exclude from work area if water >4°C. Replace lost habitat	Not covered	Not mentioned	Not covered	None	Not mentioned
Comments	Crayfish not in SGN			No frequency or limitations	Each has biases; affected by water temperature, season, flow and food	No guidance or reference								

Species/ group	Introduction to SA initial Table (1) species	Timing SA species detailed text: times	Timing SGN times	Field methods SA field methods	Field methods Standard reference	Field methods SGN field methods	Impacts SA impact factors list	Impacts SGN impact lists	Mitigation SA mitigation method	Mitigation SGN mitigation methods	Post-development monitoring SA	Post-development monitoring SGN	References SA	References SGN
Protected plants 23 4 2015	Not in Table 1	No details	Mid-April to mid-September	Survey if records suggest possibility of occurrence. Refers to missing CIEEM guidance link. Small areas: survey and make inventory. Larger areas: targeted habitat survey	Techniques and times depend on species	Do phase 2 if there is interest according to desk or preliminary survey	Change ground water regime. Altering soil conditions. Killing/damaging plants	None	Mitigation hierarchy: limit scale of works. Minimise site traffic. Create new habitat areas. Last resort: move plants	Mitigation hierarchy: Avoid area. Reduce impacts on area, reduce site traffic; improve habitats; create new habitats; enhance habitats with local species. Keep/make areas as large as possible	No mention	No mention	None	SGN
Comments	Desk study to be used for records	No reference to desk study		CIEEM web link broken	Of little LPA value	Too general to help an LPA	Too general for use; lag element ignored. No size or scalar. Fragmentation/connectivity ignored			No buffer distances given or connectivity or timescales	A basic gap for LPA use	A basic gap for LPA use	None	Link to CIEEM brings up error terms; not suitable for direct use

Species/group	Timing			Field methods			Impacts		Mitigation		Post-development monitoring		References
	Introduction to SA initial Table (1) species	SA species detailed text: times	SGN times	SA field methods	Standard reference	SGN field methods	SA impact factors list	SGN impact lists	SA mitigation method	SGN mitigation methods	SA	SGN	SA	SGN
Freshwater fish 28 3 2015	Not in Table 1	Not mentioned	Not covered	Bankside counts. Underwater counts. Electro-fishing. Seine net trawl	Bankside counts. Underwater counts. Electro-fishing. Seine net trawl. Throw nets. Hook and line Gill net	Not covered	Silt increase. Block migration. Excess light or noise. Water quality change	Not covered	Mitigation hierarchy: change timetable. Change layout. Retain connectivity. Avoid sediment ingress. Avoid light and noise	Not covered	Not mentioned	Not covered	None	Not mentioned
Comments				No frequency or limitations cited	Each has range of biases (e.g., affected by water temperature, season, flow		Scalars ignored		Scalars and spacing ignored					

Species/group	Timing		Field methods			Impacts		Mitigation		Post-development monitoring		References	
	Introduction to SA Initial Table (1) species	SA species detailed text: times	SGN times	SA field methods	Standard reference	SGN field methods	SA impact factors list	SGN impact lists	SA mitigation method	SGN mitigation methods	SA	SGN	SA	SGN
Ancient woodland and veteran trees	Not in Table 1	No details	Not mentioned in SGN	No formal entry. Reference is to inventory entries and confirmatory fieldwork. LPA to ask for tree (BS 5837) and ecological survey – presume PEA	Effectively the same as SA	Not mentioned	Direct: Part/whole loss. Root/understorey damage. Soil compaction. Pollute around tree. Change water table. Indirect: Break connectivity. Reduced seminatural habitat around woodland. Increased pollution-including dust. Increased disturbance and access	Not mentioned	Mitigation hierarchy: limit scale of works and pollution. Keep open spaces by trees. Institute buffer. Minimise site traffic. Create new habitat areas. Last resort: move plants	Mitigation hierarchy: avoid area; reduce impacts on area, reduce site traffic; reduce disturbance; change paths; institute buffer zones. Compensation not suited for ancient trees; tree planting at best is only part compensation	No mention	No mention	Two web links	

5 11 2018

Species/group	Timing			Field methods			Impacts		Mitigation		Post-development monitoring		References
	Introduction to SA initial Table (1) species	SA species detailed text: times	SGN times	SA field methods	Standard reference	SGN field methods	SA impact factors list	SGN impact lists	SA mitigation method	SGN mitigation methods	SA	SGN	SA	SGN
Comments				Web links for inventories			Requires buffer zone, cites details				A basic gap for LPA use			
	Desk study to be used for records of trees													

NE= Natural England; SA = standing advice; PBP= Partnership for Biodiversity Planning; WAC = Wildlife Assessment Check; SGN = Species Guidance Note; BCT = Bat Conservation Trust; PEA = Preliminary Ecological Assessment; LPA = local planning authority; DLL = district level licensing; CIEEM = Chartered Institute of Ecological and Environmental Management; NBN = National Biodiversity Network

References and further reading

ALGE (2007) *Validation of planning applications. Template for biodiversity and geological conservation.* London: ALGE.

ALGE (2013) *Ecological capacity and competence in English planning authorities. What is needed to deliver statutory obligations for biodiversity?* London: ALGE.

ALGE (2016a) *Evidence Submitted by ALGE to the Environmental Audi Committee Inquiry into the Future of the Natural Environment After the EU Referendum.* London: ALGE.

ALGE (2016b) *ALGE Survey: Ecological Reports – Are they Fit for Purpose? Summary of Findings.* London: ALGE.

ALGE (2020) *Implications for Local Government of delivering the Environment Bill and the Government's 25-year plan to improve the environment.* London: ALGE.

ARG (2011) *Great Crested Newt habitat suitability index.* www.arg.uk.org

Abrahams, C. (2019) Ecological survey requirements: conflicts between local validation checklists and national guidance. *In Practice* 103: 10–12.

Abrahams, C., and Nash, D.J. (2018) Do we need more evidence-based survey guidance? *In Practice* 100: 53–6.

Adams, A.M., Jantzen, M.K., Hamilton, R.M. and Fenton, M.B. (2012) Do you hear what I hear? Implications of detector selection for acoustic monitoring of bats. *Methods in Ecology and Evolution* 3: 992–8.

Andrews, H. (2018) *Bat Roosts in Trees.* Exeter: Pelagic.

Andrews, H. and Pearson, L. (2022) *Review of empirical data in respect of emergence and return times reported for the UK's native bat species version 6.* https://drive.google.com/file/d/1DeGHxyr9-p5XH6R6CRimsmquVD188WY8.

ARG (2010) *Great Crested Newt habitat suitability index.* ARG UK advice note no 5.

BCT (2022) *Interim guidance note: use of night vision aids for bat emergence surveys and further comment on dawn surveys.* London: Bat Conservation Trust.

BSI (2012) *BS 5837 Trees in relation to design , demolition and construction.* London: British Standards Institute.

BSI (2013) *BS 42020 Biodiversity – code of practice for planning and development.* London: British Standards Institute.

BSI (2021) *BS 8683: Process for designing and implementing biodiversity net gain – specification.* London: British Standards Institute.

BTO (2020) *Breeding Bird Survey* British Trust for Ornithology. *https://www.bto.org* › our-science › projects› breeding.

REFERENCES AND FURTHER READING | 233

Balestrieri, A., Remonti, L. and Prignoni, C. (2011) Detectability of the Eurasian Otter by standard surveys: an approach using marking intensity to estimate false negative rates. *Naturwissenschaften* 98: 23–31.

Baltazar-Soares, M., Pinder, A.C., Harrison, A.J., Oliver, W., Picken, K., Britton, R.J. and Andreou, D. (2022) A non-invasive eDNA tool for detecting sea lamprey larvae in river sediments: Analytical validation and field testing in a low-abundance ecosystem. *Journal of Fish Biology* 100: 1455–63.

Beebee, T. and Denton, J. (1997) *The Natterjack Toad Conservation Handbook*. Peterborough: Natural England.

Bennett, A., Ratcliffe, P., Jones, E., Mansfield, H. and Sands, R. (2005) 'Other mammals', in: D. Hill, M. Fasham, G. Tucker, M. Shewry and P. Shaw (eds) (2005) *Handbook of Biodiversity Methods*. Cambridge: Cambridge University Press.

Berthinussen, A. and J., Altringham (2012) Do Bat Gantries and Underpasses Help Bats Cross Roads Safely? *PLoS One*. https://doi.org/10.1371/journal.pone.0038775

Bibby, C.J., Burgess, N.D., Hill, D.A. and Mustoe, S.H. (2002) *Bird Census Techniques*. 2nd edition. London: Academic Press.

Bird Survey and Assessment Group (2023). Bird survey guidelines for assessing ecological impacts. V 1.1.1. https://birdsurveyguidelines.org/

Blake, W. (1808) *Milton: A Poem in Two Books*. London. Keynes Blake.

Boulton, I., Dodd, M., Hootton, S., Oxford, M. and Waymont, S. (2021) Why effective ecological reports are essential. *In Practice* 111: 13–19.

Bright, P., Morris, P. and Mitchell-Jones, T. (2006) *The Dormouse Conservation Handbook*. Peterborough: Natural England.

Brown, A.F. and Shepherd, K. (1993) A method for censusing upland breeding waders. *Bird Study* 40: 89–95.

Buglife (2019) *Good Planning Practice for Invertebrate Surveys*. Peterborough: Buglife.

Burns, F., Mordue, S., al Fulaij, N., Boersch-Supan, P.H., Boswell, J., Boyd, R.J., Bradfer-Lawrence, T., de Ornellas, P., de Palma, A., de Zylva, P., Dennis, E.B., Foster, S., Gilbert, G., Halliwell, L., Hawkins, K., Haysom, K.A., Holland, M.M., Hughes, J., Jackson, A.C., Mancini, F., Mathews, F., McQuatters-Gollop, A., Noble, D.G., O'Brien, D., Pescott, O.L., Purvis, A., Simkin, J., Smith, A., Stanbury, A.J., Villemot, J., Walker, K.J., Walton, P., Webb, T.J., Williams, J., Wilson, R., Gregory, R.D. (2023). *State of Nature 2023*. The State of Nature partnership. www.stateofnature.org.uk.

Calladine, J., Garner, G., Wernham, C. and Thiel, A. (2009) The influence of survey frequency on population estimates of moorland breeding birds. *Bird Study* 56: 381–8.

Camden Borough Council (2022) *Biodiversity in Camden*. London: CBC.

Carrington, D. (2019) Widespread losses of pollinating insects revealed across Britain. https://www.theguardian.com/environment/2019/mar/26/.

Carver, L. and Sullivan, S. (2017) How economic contexts shape calculations of yield in biodiversity offsetting. *Conservation Biology* 31: 1053–65.

Chanin, P. (2003a) *Ecology of the European Otter*. Peterborough: English Nature.

Chanin, P. (2003b) *Monitoring the Otter Lutra lutra*. Peterborough: English Nature.

Chanin, P. (2005). *Otter surveillance in SACs: testing the protocol. English* Nature Research Report No. 664. Peterborough: English Nature.

Chanin, P. and Gubert, L. (2011) Surveying hazel dormice (Muscardinus avellanarius) with tubes and boxes: a comparison. *Mammal Notes*: summer 1-6.

Cherrill, A. (2013) Repeatability of vegetation mapping using phase 1 and NVC approaches. *In Practice* 81: 41–5.

Cherrill, A. (2016) Inter-observer variation in habitat survey data: investigating the consequences of professional practice. *Journal of Environmental Planning and Management* 59: 1813–32.

Chelmsford District Council. (2009) *Biodiversity Check List*. Chelmsford D.C., Chelmsford.

Chichester District Council. (2018) *Guidance on Ecological Surveys and Planning Applications*. Chelmsford D.C., Chelmsford. https://www.chichester.gov.uk/media/8167/Guidance-on-ecological-surveys-and-planning-applications/pdf/Guidance_on_Ecological_Surveys_and_Planning_Applications_March_2018.pdf

CIEEM (2013) *Code of Professional Conduct*. Winchester: CIEEM.

CIEEM (2013a) *Competencies for Species Survey: Great Crested Newt*. Winchester: CIEEM.

CIEEM (2013b) *Competencies for Species Survey: Hazel Dormouse*. Winchester: CIEEM.

CIEEM (2013c) *Competencies for Species Survey: White-clawed Crayfish*. Winchester: CIEEM.

CIEEM (2013d) *Competencies for Species Survey: Bats*. Winchester: CIEEM.

CIEEM (2013e) *Competencies for Species Survey: Otter*. Winchester: CIEEM.

CIEEM (2013f) *Competencies for Species Survey: Badger*. Winchester: CIEEM.

CIEEM (2013g) *Competencies for Species Survey: Natterjack toad*. Winchester: CIEEM.

CIEEM (2013h) *Competencies for Species Survey: White Clawed Crayfish Bats*. Winchester: CIEEM.

CIEEM (2014) *Competencies for Species Survey: Reptiles*. Winchester: CIEEM

CIEEM (2017) *Guidelines for Preliminary Ecological Appraisal, 2nd edition*. Winchester: CIEEM.

CIEEM (2018) Evidence to House of Lords review: The countryside at a crossroads: Is the Natural Environment and Rural Communities Act 2006 still fit for purpose? https://publications.parliament.uk/pa/ld201719/ldselect/ldnerc/99/9908.htm

CIEEM (2019) *Competency Framework*. Winchester: CIEEM.

CIEEM (2019a) *Guidelines for Ecological Impact Assessment in the UK and Ireland*. Winchester: CIEEM.

CIEEM (2019b) *On the Lifespan of Ecological Reports and Surveys*. Winchester: CIEEM.

CIEEM (2020) *Guide on ecological survey and assessment in the UK during the covid outbreak. Version 3.* Winchester: CIEEM.

CIEEM (2021a) *Good practice guidance for habitats and species*. Winchester: CIEEM.

CIEEM (2021b) *Competency Standard for Water Vole Survey, Mitigation and Management*. Winchester: CIEEM.

CIEEM (2021c) *Competency Standard for Badger Survey, Mitigation and Management*. Winchester: CIEEM.

CIEEM (2021d) *Competency Standard for Natterjack Toad Survey, Mitigation and Management*. Winchester: CIEEM.

CIEEM (2021e) *Competency Standard for Otter Survey, Mitigation and Management.* Winchester: CIEEM.

CIEEM (2022) *Scottish Local Planning Authority Ecological Expertise and Capacity Survey Report March 2022.* Winchester: CIEEM.

CIEEM (2022a) *Code of professional conduct.* Winchester: CIEEM.

CIEEM (2022b) *Competency Standard for Reptile Survey, Mitigation and Management.* Winchester: CIEEM.

Cocker, M. (2018) *Our Place.* London: Jonathan Cape.

Collins, J. (ed.) (2016) *Bat Surveys for Professional Ecologists.* London: Bat Conservation Trust.

Crawley, D., Coomber, F. Kubasiewicz, L., Harrower, C., Evans, P., Waggitt, J., Smith, B. and Mathews, F. (2020) *Atlas of the Mammals of Great Britain and Northern Ireland.* Exeter: Pelagic Publishing.

Crosher, I., Gold, S., Heaver, M., Heydon, M., Moore, L., Panks, S., Scott, S., Stone, D. and White, N. (2019) *The Biodiversity Metric 2.0: auditing and accounting for biodiversity value. User guide (Beta Version, July 2019).* Peterborough: Natural England.

DEFRA (July 2013) *Outputs from: Managing Risk in Habitats Regulations Assessment. A workshop for Local Planning Authorities.* Organised by DEFRA, CIEEM, ALGE and the POS. DEFRA: London.

DEFRA (August 2016) *England Natural Environment Indicators.* https://assets.publishing.service.gov.uk/media/5a81c92f40f0b62305b90cc0/ENEI_16_final_Revised.pdf

DEFRA. (2021) Environment Act 2021. https://bills.parliament.uk/bills/2593.

DEFRA. (2023) Local nature recovery strategies. https://www.gov.uk/government/publications/local-nature-recovery-strategies/local-nature-recovery-strategies.

Dean, M., Strachan, R., Gow, D. and Andrews, R. (2016) *The Water Vole Mitigation Handbook.* Southampton: The Mammal Society.

Dean, M. (2021) *Water Vole Field Signs and Habitat Assessment: A Practical Guide to Water Vole Surveys.* Exeter: Pelagic Publishing.

Dean, M., Edmonds, B. and Downey, H. (2021) Good practice guidance: where's the evidence? *In Practice* 112: 51–4.

Delahay, R., Wilson, G., Harris, S. and Macdonald, D. (2008) In: S. Harris and D. Yalden D (eds). *Mammals of the British Isles: Handbook,* 4th Edition. Southampton: The Mammal Society.

Dobson, F.S. (1992) *Lichens.* Slough: Richmond Publishing.

Dorset Council. (2023) *Ecology survey guide.* Dorchester: Dorset Council.

Drake, C.M., Loot, D., Alexander, K. and Webb, J. (2007) Surveying terrestrial and freshwater invertebrates for conservation evaluation. NE Research Report NERR005.

Drayton, K. and Thompson, S. (2013) Ecological mitigation measures in English Impact Assessment. *Journal of Environmental Management* 119: 103–10.

EC LIFE. Sturgeon. https://webgate.ec.europa.eu/life/publicWebsite/index.cfm?fuseaction=search.dspPageandn_proj_id=495

Edgar P. and Bird, D.R. (2006) *Action plan for the conservation of the sand lizard* (Lacerta agilis) *in Northwest Europe.* Council of Europe Standing Committee November 2006.

Edgar, D., Foster, J. and Baker, J. (2010) *Reptile Habitat Management Handbook*. Bournemouth: Amphibian and Reptile Conservation.

Elgar, E. (1922) *Jerusalem*. London: Goodmusic.

ENDS (2019) Capacity crunch: do councils have the expertise to deliver their biodiversity goals? *ENDS Report* May 2019.

Environment Agency. (2019) Sampling eel populations in rivers. *Operational instruction 778-06*. Bristol: Environment Agency.

Environment Agency (2022) *Working with Nature*. Bristol: Environment Agency.

Essex County Council (2015) *Essex Biodiversity validation checklist* v 1.3. Chelmsford: ECC.

Fay, N. (2007) Defining and surveying veteran and ancient trees. *Working paper. March 2007*. Treework Environmental Practice. www.treeworks.co.uk.

Fay, N. and de Berker, N. (1997) *The Specialist Survey Method*. Veteran Trees Initiative. Peterborough: Natural England.

Fay, N. and Rose, B. (2004) Survey methods and development of innovative arboricultural techniques. https://citeseerx.ist.psu.edu/document?repid=rep1andtype=pdfanddoi=23cef04a6158ae4da088c71e58dc52355e414fd4. Flowers, H.J. (2013) A novel approach to surveying sturgeon using side-scan sonar and occupancy modelling. Marine And Coastal Fisheries: Dynamics, Management, and Ecosystem Science 5: 211–23. Forestry Commission and Natural England. (2013) *Guidance on Managing Woodlands with Sand Lizard and Smooth Snake in England*. Peterborough: Natural England.

Francois, D., Ursenbacher, S., Bossinot, A., Ysnel, F. and Lourdais, O. (2021) Isolation by distance and male-biased dispersal at a fine spatial scale: a study of the common European adder (*Vipera berus*) in a rural landscape. *Conservation Genetics* 22: 823–37.

Froglife. (1999) *Reptile Survey*. Froglife Advice Sheet No. 10. Peterborough: Froglife.

Froidevaux, J.S.P., Boughey, K.L., Hawkins, C.L., Jones, G. and Collins, J. (2020) Evaluating survey methods for bat roost detection in ecological impact assessment. *Animal Conservation* 23: 597–606.

Gent, T. and Gibson, S. (2003) *Herpetofauna Workers' Manual*. Peterborough: JNCC.

Gilbert, F. (2000). *Lichens*. London: Collins.

Gilbert, G., Gibbons, D.W. and Evans, J. (1998) *Bird Monitoring Methods*. Sandy: RSPB.

Gillings, S., A.M., Wilson, A.M., Conway, G.J., Vickery, J.A., Fuller, R.J., Beavan, P., Newson, S., Noble, D. and Toms, M.P. (2008) *Winter Farmland Bird Survey*. BTO Research Report No. 494. Thetford: BTO.

Gove, M. (2018) *A Green Future: Our 25 Year Plan to Improve the Environment*. https://assets.publishing.service.gov.uk/government/uploads/system/uploads/attachment_data/file/693158/25-year-environment-plan.pdf.

HMG. (1992) *Protection of Badgers Act 1992*. https://www.legislation.gov.uk/ukpga/1992/51/contents.

HMG. (2006) Natural Environment and Rural Communities (NERC) Act 2006. https://www.legislation.gov.uk/ukpga/2006/16/contents

HMG. (2021) National Planning Policy Framework (NPPF). https://www.gov.uk/government/publications/national-planning-policy-framework--2

HMG (2022) Biodiversity mitigation plan checklist. https://assets.publishing.service.gov.uk/government/uploads/system/uploads/attachment_data/file/934786/biodiversity-mitigation-plan-checklist.pdf

Hardey, J., Crick, H., Wernham, C., Riley, H., Etheridge, B. and Thompson, D. (2013) *Raptors*. Edinburgh: TSO.

Harper, L., Lawson Handley, L., and Carpenter, A. (2019) Environmental DNA (eDNA) metabarcoding of pond water as a tool to survey conservation and management priority mammals. *Biological Conservation*. ISSN 0006-3207 https://doi.org/10.1016/j.biocon.2019.108225

Harris, S., Cresswell, P. and Jeffries, D. (1989) *Surveying Badgers*. London: The Mammal Society.

Hayhow, D.B., Eaton, M.A., Stanbury, A.J., Burns, F., Kirby, W.B., Bailey, N., Beckmann, B., Bedford, J., Boersch-Supan, P.H., Coomber, F., Dennis, E.B., Dolman, S.J., Dunn. E., Hall, J., Harrower, C., Hatfield, J.H., Hawley, J., Haysom, K., Hughes, J., Johns, D.G., Mathews, F., McQuatters-Gollop, A., Noble, D.G., Outhwaite, C.L., Pearce-Higgins, J.W., Pescott, O.L., Powney, G.D. and Symes, N. (2019) *The State of Nature*. The State of Nature partnership.

Hearn, S.M., Healey, J.R., McDonald, M.A., Turner, A.J., Wong, J.L.G., and Stewart, G. (2011) The repeatability of vegetation classification and mapping. *Journal of Environmental Management* 92: 1174–84.

Helm, D. (2019) *Green and Prosperous Land: A Blueprint for Rescuing the British Countryside*. London: Collins.

Highland Regional Council. Best practice guidance- model Badger protection plan (BPP) (2006). https://www.highland.gov.uk/download/downloads/id/2637/Badger_best_practice_guidance_Badger_surveys_september_2006.pdf

Hill, D., Fasham, M., Tucker, G., Shewry, M. and Shaw, P. (2005) *Handbook of Biodiversity Methods*. Cambridge: Cambridge University Press.

Hillman, R.J. (2000) *Monitoring Allis and Twaite Shad*. Peterborough: Natural England.

Hillman, R.J., Cowx, I.G. and Harvey J.P. (2003) *A standardised survey and monitoring protocol for the assessment of shad populations within SAC rivers*. Conserving Natura 2000 Rivers Monitoring Series No. 3. Peterborough: English Nature. http://www.english-nature.org.uk/lifeinukrivers/publications/shad monitoring.pdf

Horner, C. and Davidson, N. (2020) 'Accounting for biodiverse wildlife corridor plantations', *Meditari Accountancy Research*. Emerald Publishing Limited.

Hounsome, T. (2021) *Bird Survey Guidelines: Why We Need Them*. https://birdsurveyguidelines.org/.

House of Commons Library. (2019) Planning enforcement in England. *Briefing Paper* 1579.

HoL (2018). *The Countryside at a Crossroads*. https://publications.parliament.uk/pa/ld201719/ldselect/ldnerc/99/9902.pd.

Hunter, S.B., zu Ermgassen, S., Downey, H., Griffiths, R.A. and Howe, C. (2021) Evidence shortfalls in the recommendations and guidance underpinning ecological mitigation for infrastructure developments. *Ecological Evidence and Solutions* 2: e12089.

Huntingdonshire District Council. (2019) Huntingdonshire's Local Plan to 2036. https://www.huntingdonshire.gov.uk/media/3872/190516-final-adopted-local-plan-to-2036.pdf.

JNCC. (2010) *Handbook for phase 1 habitat survey*. Peterborough: JNCC.

JNCC. (2015) *Commons standards monitoring guidance for freshwater fauna*. Peterborough, JNCC.

Juskaitis, R. and Buchner, S. (2013). *The Hazel Dormouse*. Hohenwarslebern: Westarp.

Klein, Z., Quist, M., Rhea, D. and Senecal, A. (2015) Sampling techniques for burbot in a western non-wadeable river. *Fisheries Management and Ecology* 22: 213–23.

Kohler, M. (2021) Early career biologists: time to root out exploitation in the Consultancy Sector. *In Practice* 111: 27–8.

Langton, T., Beckett, C. and Foster, J. (2001) *Great Crested Newt Conservation Handbook*. Halesworth: Froglife.

Lewis, B., Griffiths, R. A., and Wilkinson, J. W. (2016) Population status of great crested newts (*Triturus cristatus*) at sites subjected to development mitigation. *Herpetological Journal* 27: 133–142.

Lintott, P.R. and Mathews, F. (2017) Basic mathematical errors may make ecological assessments unreliable. *Biodiversity and Conservation* 27: 265–7.

Lintott, P. and Mathews, F. (2018) Reviewing the evidence on mitigation strategies for bats in buildings informing best-practice for policy makers and practitioners. https://uwe-repository.worktribe.com/output/868713.

Macdonald, D.W. and Tattersall, F. (2001) *Britain's Mammals: The Challenge for Conservation*. London: Peoples Trust for Endangered Species.

Mammal Society. (1963) Badger survey. In S. Harris S and D. Yalden D. (eds) 2008. *Mammals of the British Isles: Handbook*, 4th Edition. Southampton: The Mammal Society.

Maitland, P.S. and Campbell, R.N. (1992) *Freshwater Fishes*. London: Collins.

Mammal Society (2023) Response to state of nature report 2023. https://www.mammal.org.uk/2023/09/state-of-nature-2023-response/

Marsh, R. (2022). Burdens not gain: have we all missed a trick? *In Practice:* 117: 52–4.

Mathews, F., Kubasiewicz, L.M., Gurnell, J., Harrower, C., McDonald R.A., and Shore, R.F. 2018. *A review of the population and conservation status of British Mammals*. A report by The Mammal Society under contract to Natural England, Natural Resources Wales and Scottish Natural Heritage. Peterborough: Natural England.

Nash, D.J., Humphries, N., and Griffiths, R.A. (2020) Effectiveness of translocation in mitigating reptile-development conflict on the UK. *Conservation Evidence* 17: 7–11.

National Biodiversity Network (2022). https://nbn.org.uk/.

Natural England. MAGIC. https://magic.DEFRA.gov.uk/.

Natural England. (2001) *Great Crested Newt Mitigation Guidelines*. Peterborough: Natural England.

Natural England. (2005) *Organising Surveys to Determine Site Quality for Invertebrates*. Peterborough: Natural England.

Natural England. (2010) Assessing the effects of onshore wind farms on birds. *Technical Information Note TIN069*. Peterborough: Natural England.

Natural England. (2009) Guidance on 'current use' in the definition of a badger sett. http://www.naturalengland.org.uk/Images/WMLG17_tcm6-11815.pdf.

Natural England. (2011a) *Reptile Mitigation Guidelines*. NE Technical Note TIN102.

Natural England. (2011b) *Badgers and Development*. Peterborough: Natural England.

Natural England (2014, 2022) Protected species and development: advice for local planning authorities. https://www.gov.uk/guidance/protected-species-how-to-review-planning-applications

Natural England. (2015) *Badgers: surveys and mitigation for development projects.* Peterborough: Natural England.

Natural England. (2015a) *Reptiles: surveys and mitigation for development projects.* Peterborough: Natural England.

Natural England. (2015b, 2020) *Bats: surveys and mitigation for development projects.* Peterborough: Natural England.

Natural England (2015c) *Hazel Dormice: surveys and mitigation for development projects.* Peterborough: Natural England.

Natural England. (2022) *Protected species and development: advice for local planning authorities.* Peterborough: Natural England.

Natural England (2022a) *Prepare a planning proposal to avoid harm or disturbance to protected species.* Peterborough: Natural England.

Natural England (2022b) *Guidance Badgers: advice for making planning decisions.* Peterborough: Natural England

Natural England (2022c) *Guidance Hazel dormice: advice for making planning decisions.* Peterborough: Natural England.

Natural England (2022d). *Guidance Freshwater Pearl Mussels*: a*dvice for making planning decisions.* Peterborough: Natural England.

Natural England (2022e) *Guidance Water Vole: advice for making planning decisions.* Peterborough: Natural England

Natural England (2022f) *Guidance Otter: advice for making planning decisions.* Peterborough: Natural England

Natural England (2022g) *Guidance Reptiles: advice for making planning decisions.* Peterborough: Natural England.

Natural England (2022h) *Guidance White Clawed Crayfish: advice for making planning decisions.* Peterborough: Natural England

Natural England (2022i) *Guidance Fish: advice for making planning decisions.* Peterborough: Natural England

Natural England (2022j) *Guidance Ancient woodland, ancient trees and veteran trees: advice for making planning decisions.* Peterborough: Natural England

Natural England (2022k) *Guidance Protected plants, fungi and lichens: advice for making planning decisions.* Peterborough: Natural England

Natural England (2022l) *Guidance Wild birds: advice for making planning decisions.* Peterborough: Natural England

Natural England (2022m) *Guidance Invertebrates: advice for making planning decisions.* Peterborough: Natural England

Natural England (2022n) *Guidance Bats: advice for making planning decisions.* Peterborough: Natural England

Natural England (2022o) *Guidance Great Crested Newts: advice for making planning decisions.* Peterborough: Natural England.

Natural England (2023) *The Biodiversity Metric. 4.0 User Guide.* Peterborough: Natural England.

Natural History Museum. (2020) UK has led the world in destroying the natural environment. https://www.nhm.ac.uk/discover/news/2020/september/uk-has-led-the-world-in-destroying-the-natural-environment.html.

NatureScot (2020) *Species and planning guidance.* Edinburgh: Nature Scot. https://www.nature.scot/professional-advice/planning-and-development/planning-and-development-advice/planning-and-development-standing-advice-and-guidance-documents

NatureScot (2020a) *Standing Advice for Planning Applications.* Edinburgh: NatureScot.

NatureScot (2020b) Standing advice for planning consultations - Badgers. Edinburgh: NatureScot. https://www.nature.scot/doc/standing-advice-planning-consultations-badgers

NatureScot (2021) *Bats and Onshore Wind Turbines: Survey, Assessment and Mitigation.* Edinburgh: NatureScot.

Neal, E. and Cheeseman, C. (1996) *Badgers.* Poyser Natural History. London: T and AD Poyser.

North Lincolnshire Council (2020). *North Lincolnshire Local Plan pt 9. Biodiversity and geodiversity.* Scunthorpe: NLC.

Northern Ireland Government (2015) *The Strategic Planning Policy Statement.* https://www.infrastructure-ni.gov.uk/publications/strategic-planning-policy-statement.

Oxford, M. (2012) Councils in adversity –why less isn't more for nature. *ECOS* 33: 21–3.

Panks, S., A, White, N., Newsome, A., Nash, M., Mayhew, E., Alvarez, M., Russell, T, Cashon, C., Goddard, F., Scott, S., Heaver, M., Butcher, B. and Stone, D. (2022) *Biodiversity metric 3.1: Auditing and accounting for biodiversity – User Guide.* Peterborough: Natural England.

Parry, C.H. (1916) *Jerusalem.* Choral song. London: Goodmusic.

Partnership for Biodiversity in Planning (PBP) (2021). Wildlife Assessment Check. https://www.biodiversityinplanning.org/.

Peay, S. (2003) *Monitoring the White-clawed Crayfish.* Peterborough: Natural England.

Peay, S. (2004) A cost-led evaluation of survey methods for white-clawed crayfish: lessons from the UK. *Bull Fr Peche Piscic* 335–52.

Planning Portal (2023) What are material considerations? https://www.planningportal.co.uk/services/help/faq/planning/about-the-planning-system/what-are-material-considerations

Plummer, K. (2022) Making space for birds. *Bird Study* 343: 12–15.

Pritchard, E., Chadwick, D., Chadwick, M., Bradley, P., Sayer, C. and Axmacher, C. (2021) Assessing methods to improve benthic fish sampling in a stony freshwater stream. *Ecological Solutions and Evidence* 2: e12111.

RTPI. (2019) *Biodiversity in Planning.* London: Royal Town Planning Institute.

Rackham, O. (1986) *The History of the Countryside.* London: Dent.

Rampling, E.E., zu Ermgassen, S.O. Hawkins, I. and Bull, J.W. (2023) Achieving biodiversity net gain by addressing governance gaps underpinning ecological compensation policies. *Conservation Biology.* https://doi.org/10.31219/osf.io/avrhf.

Reed, T.M. (2019) Baselines – knowing where you start and how you are progressing. *Conservation Land Management* 17: 29–34.

Reed, T.M. (2019) Transparency, open evaluation and the use of Professional judgement in planning applications. *In Practice* 103: 13–17.

Reed, T.M. (2020) Does the NE/DEFRA standing advice for protected species tell local planning authorities what to do? Is it fit-for-purpose? *Town and Country Planning* 89: 62–70.

Reed, T.M. (2021) Limitations sections in UK impact assessments, and how to fix them. *In Practice* 111: 20–3.

Reynolds, P. and Harris, M. (2005) Inverness badger survey 2003. *SNH Commissioned Report No 096*. Edinburgh: Scottish Natural Heritage.

Robertson, M. (2021) The state of no net loss/net gain and biodiversity offsetting policy in English local planning authorities: full report. https://cieem.net/resource/lpa-survey-morgan-robertson/.

Russ, J. (2012). *British Bat Calls: A Guide to Species Identification*. Exeter: Pelagic.

Scarborough Borough Council. (2004) Biodiversity Action Plan. Scarborough: Scarborough Borough Council. https://www.scarborough.gov.uk/biodiversity-action-plan.

Scottish Government (2016) *National Planning Framework*.

Scottish Government (2020) Scottish biodiversity strategy post-2020: statement of intent. https://www.gov.scot/publications/scottish-biodiversity-strategy-post-2020-statement-intent/

SNH. (2017) Scottish Natural Heritage Recommended bird survey methods to inform impact assessment of onshore wind farms. https://www.nature.scot/sites/default/files/2018-06/Guidance%20Note%20-%20Recommended%20bird%20survey%20methods%20to%20inform%20impact%20assessment%20of%20onshore%20windfarms.pdf.

Scottish Badgers. (2018) *Surveying for badgers: good practice guidelines*. https://www.scottish-badgers.org.uk/wp-content/uploads/2020/12/Surveying-for-Badgers-Good-Practice-Guidelines_V1-2020-2455979.pdf.

Sewell, D., Griffiths, R.A., Beebee, T., Foster, J. and Wilkinson, J.W. (2013) *Survey protocols for the British herpetofauna*. University of Kent: DICE.

Sheail, J. (1998) *Nature Conservation in Britain: The Formative Years*. London: The Stationery Office.

Sheldrake, M. (2020) *Entangled Life*. London: Bodley Head.

Singh, G.G., Lerner, J., and Mach, M. (2020) Scientific shortcomings in environmental impact statements internationally. *People and Nature* 2: 369–79.

Sinsch, U., Oromi, N., Miaud, C., Denton, J. and Sanuy, D. (2012) Connectivity of local amphibian populations: Modelling the migratory capacity of radio-tracked Natterjack Toads. *Animal Conservation* 15: 388–96.

Snell, L. and Oxford, M. (2022) *Survey of local planning authorities and their ability to deliver biodiversity net gain in England. Do LPAs currently have the necessary expertise and capacity?* London: ALGE.

Stone, E.L., Jones, G., and Harris, S. (2013). Mitigating the effect of development on bats in England with derogation licensing. *Conservation Biology* 27: 1324–1334.

Strachan, R., Moorhouse, T. and Gelling, M. (2011) *Water Vole conservation handbook*. Oxford: Wildcru.

Suffolk Biodiversity Information Service (2021). *Sandy stiltball*. Ipswich: SBIS

Tang, C.Q. (2020) *eDNA monitoring for migratory fish assemblages*. NECR 290. Peterborough: Natural England.

The Wildlife Trusts. (2023) *HS2 Double Jeopardy How the UK's Largest Infrastructure Project Undervalued Nature and Overvalued Its Compensation Measures*. https://www.wildlifetrusts.org/sites/default/files/2023-02/23JAN_HS2_Double_Jeopardy_FINAL01.02.23.pdf

Thompson, D., Graves, R., Hayns, S. and Alexander, D. 2016. The alternative Decalogue for CIEEM members: towards improving standards in the profession. *In Practice* 91: 63–6.

Tree, I. (2018) *Wilding*. London: Picador.

Treweek, J. and Thompson, S. (1997) A review of ecological mitigation measures in UK environmental statements with respect to sustainable development. *International Journal of Sustainable Development and World Ecology* 4: 40–50.

Tyldesley D. and Bradford G. (2012) *Planning Policy and Biodiversity Offsets Report on Phase II Research: Effectiveness of the Application of Current Planning Policy in the Town and Country Planning System*. Report to the Department of Environment, Food and Rural Affairs.

UK Habitat (2020) The UK habitat classification user manual. https://ukhab.org/ukhab-documentation/

Welsh Government. (2016) Environment (Wales) Act 2016. https://www.legislation.gov.uk/anaw/2016/3/contents/enacted.

Woodfield, D. (2021) Development and net gain: promise and reality. *In Practice* 111: 46–9.

Woodland Trust (2016). Comments on Smithy Wood. https://www.woodlandtrust.org.uk/protecting-trees-and-woods/campaign-with-us/smithy-wood/

Young, M., Hastie, L.C. and Cooksley, S.L. (2003) *Monitoring the Freshwater Pearl Mussel*. Peterborough: English Nature.

Zu Ermgassen, S. and Bull, J. (2020) Will biodiversity net gain improve English biodiversity? Results from the first evaluation of net gain, and what's next. https://www.wcl.org.uk/will-biodiversity.

Zu Ermgassen, S. (2021) The biodiversity metric 3.0. *British Wildlife* 33: 73.

Zu Ermgassen, S., Marsh, S., Ryland, K., Church, E., Marsh, R. and Bull, J. (2021) Exploring the ecological outcomes of mandatory biodiversity net gain using evidence from early-adopter jurisdictions in England. *Conservation Letters* 14: e12820.

Zu Ermgassen, S., Milner Gulland, E.J. and Bull, J. (2022) Biodiversity: why new rules to ensure nature benefits from building projects could fail. *The Conversation*: March 24. https://theconversation.com/biodiversity-why-new-rules-to-ensure-nature-benefits-from-building-projects-could-fail-179701

Zu Ermgassen, S., Milner Gulland, E.J., Addison, P., Baker, J., Bateman, I., Bull, J., Jones, J.P., Smith, B. and Treweek, J. (2022a) *An open letter to the Rt Hon Michael Gove, the Rt Hon George Eustice and Tony Juniper: Ensuring that mandatory Biodiversity Net Gain fulfils its potential for nature recovery*. https://www.oxfordmartin.ox.ac.uk/downloads/academic/BNG-Open-Letter_020322.pdf

Index

Note: Page numbers in *italics* refer to tables.

Abrahams, C. 12, 24–25
Adders *126*
Allis Shad and Twaite Shad, species guidance *143–147*
   field collection methods *145–146*
      adults *145*
      eggs *144*
      juveniles *145*
   impacts *146*
   monitoring *147*
   site details *143–144*
   survey details *145–146*
   surveys objective *144*
ancient trees and veteran trees (AVT), species guidance *151–158*
   compensation measures *157*
   direct impacts *156*
   field collection methods *152–153*
   impacts *155–156*
   pre-2022 publication *230*
   site details *151*
   survey details *154*
   surveys objective *152*
ancient woodland, species guidance *151–158*
   compensation measures *157*
   direct impacts *156*
   field collection methods *152–153*
   impacts *155–156–157*
   pre-2022 publication *230*
   site details *151*
   survey details *154*
   surveys objective *152*
Association of Local Government Ecologists (ALGE), ecology and planning 3, 7, 16–25

   capacity 18
   CIEEM Competency Framework 18
   DEFRA-sponsored study 18, 20–21
   ecological skills gap in 19
   ecological surveys 18
   environmental management 19
   impact/habitat regulation assessments 19
   professional competence 18
   scientific method 18
   self-reported limitations 20
   smaller-scale applications 17
   surveys conducted 17
   technical resilience 18

Badger, species guidance 59–60, *96–100*
   field collection methods *97–98*
      day nests *97*
      faeces *97*
      footpaths *97*
      footprints *97*
      hair traces *97*
      scratching posts *97*
      setts *97*
      snuffle holes *97*
   impacts *100*
   pre-2022 publication *217*
   site details *96*
   survey details *98–99*
   surveys objective *99*
Bat Conservation Trust (BCT) 60
bat surveys *191*
bats, species guidance 60–61, *184–196*
   field collection methods *185–192*
      bat surveys *191*
      desk study *191*

244 | PROTECTED SPECIES AND BIODIVERSITY

    medium to large areas *189*
    preliminary roost assessment (PRA) *186*
    roost characterisation survey (RCS) *188*
    static recording/automated survey *189–190*
    survey methods *191*
    swarming survey *188–189*
    transects *189–190*
    windfarms *191*
  impacts *195–196*
  interpreting the data collected *192*
  pre-2022 publication *218*
  site details *184*
  survey details *192–194*
  survey needs *187*
  survey timing *189–190*
  surveys objective *185–186*
before–after control–impact approach (BACI) 209
Bibby, C.J. 52, 175
biodiversity 5–25
  evaluation, policy needs for 5–11
  guidance for applicants and planners 26
  LPAs and, strategic approach 7–11
    CIEEM/ALGE (CIEEM 2019) checklist, *10–11*
    simplified planning flowchart (ALGE) 9
  metric principles 213
  in planning applications 18
Biodiversity Net Gain (BNG) and species 33, 71, 74, 209–212
  BNG 3.1 209–211
  BNG 4 209–210
  BNG 4.0 209
  testing BNG 212–214
Biodiversity Standard 3
Bird Survey Guidelines (BSG 2023) 51
birds, pre-2022 publication *219*
bottle traps *199*
Boulton, I. 3
Bradford, G. 17–18
breeding bird survey *169*

British biodiversity standard BS 42020 63–70, 206
  basics 63–70
  decisions based upon adequate information 68
  development proposals 64–65
  ecological information 66
  ecological judgment and advice 68
  ecological reports 66
  general 64
  identifying limitations 66
  professional judgement 65
  professionals involved 64
  reference to technical competence 64
British Trust for Ornithology (BTO) 51
BS 42020
  see British biodiversity standard BS 42020
BS 5837 153

Camden Borough Council 206
capacity 18
Chanin, P. 107, *120–123*
Chartered Institute of Ecology and Environmental Management (CIEEM) 7, 62, 206
  CIEEM Competency Framework 18
    accomplished 18
    authoritative 18
    basic 18
    capable 18
Chelmsford District Council (CDC) 12, *14*
  differing expectations and requirements of *14*
Chichester District Council (CC) 12, *14*, 25, 206
  differing expectations and requirements of *14*
Chichester protected species check list *13*
Collins, J. 32, 60–61, 89, *186*, *194*
Common Bird Census Breeding bird survey (CBC) *170*
common snakes *125*
Crime and Disorder Act 1998 6

data to the planners, providing 62–70
    see also British biodiversity standard
        BS 42020
de Berker, N. 153
Dean, M. 116
DEFRA species survey guidance 12
Department of Environment, Food and
    Rural Affairs (DEFRA) 26
district level licensing (DLL) 202, 204
Dorset County Council 206
Drake, C.M. 180
Drayton, K. 213

Ecological Impact Assessments
    (EcIA) 51
ecological mitigation and compensation
    measures (EMC) 80, 208
Edgar, D. 126
English Government's 25-year Plan 2
Environment (Wales) Act of 2016 7
Environment Bill (2021) 2
environmental DNA (eDNA) 199
Environmental Impact Assessments
    (EIA) 51
Essex County Council 207
European protected species (EPS)
    82, 147

Fay, N. 153
field data collection methods and
    reliability 89–92
    Badger 97–98
    distance 92
    light conditions 91–92
    limitations 92
    number of times to visit 91
    objectives stated 90
    survey length 91
    time of visit 91
    timing a visit 90
    weather 91
Francois, D. 125
freshwater fish, species guidance 136–150
    Allis Shad and Twaite Shad 143–147
    impacts 138
    pre-2022 publication 229

site details 136
Sturgeon 147–150
survey details 136–137
Whitefish/Vendace 139–143
Freshwater Pearl Mussel, species
    guidance, 111–114
    field collection methods 112–113
    impacts 114
    site details 111
    survey details 113
    surveys objective 112
Froglife 200
fungi, species guidance 159–167
    field collection methods 161
    impacts 166
    monitoring 167
    site details 159–160
    survey details 164–165
    surveys objective 160

Gent, T. 126–127, 199–200
Gibson, S. 126–127, 199–200
Gilbert, G. 52, 171–172
Grass Snake 125–126
Great Crested Newt (GCN), species
    guidance 197–205
    field collection methods 199–200
        bottle traps 199
        egg searches 200
        environmental DNA 199
        netting 200
        pitfall traps 199
        terrestrial searching 200
        torch survey 200
    impacts 204
    site details 197–198
    survey details 202–203
    surveys objective 198
    survey timing 201
Great Crested Newts
    mitigation actions 208
    pre-2022 publication 221
green and pleasant land 1–2
guidance and interpretation 26–61
    biodiversity guidance for applicants
        and planners 26

see also Natural England (NE) guidance; Partnership for Biodiversity in Planning (PBP)

Hardey, J. 172
Hazel Dormouse, species guidance 105–110
   field collection methods 107–108
      boxes and nest tubes 107
      hair tubes 108
      nest boxes 107
      nest tubes 107
      nests 108
      nuts 107
      trapping 108
   impacts 109–110
   pre-2022 publication 220
   site details 105
   survey details 108–109
   surveys objective 106
Hill, D. 153, 161
Hounsome, T. 51, 54
Hunter, S.B. 80–81, 208–209
Huntingdonshire District Council (HDC) 12, 14
   differing expectations and requirements of 14

impact assessment and conservation payment certificate (IACPC) 203
Individual species and group-specific best practice methods (SGN) from the PBP 26
invertebrates (IV), species guidance 178–183
   field collection methods 180–181
   impacts 182
   pre-2022 publication 222
   site details 178–179
   survey details 181
   survey timing 181
   surveys objective 179–180

Langton, T. 200
Lewis, B. 208
lichens, species guidance 159–167
   field collection methods 161–162
      colony size and extent 162
      habitat description 162
      quadrats 162
   impacts 166
   monitoring 167
   site details 159–160
   survey details 164–165
   surveys objective 160
limitations 56
   in ancient woodland, ancient trees and veteran trees species guidance 155
   in Badger species guidance 98–99
   in bats species guidance 194
   BNG 3.1 on 211
   BSG setting out 52
   in data collected and offered in applications 64, 66, 71, 80, 88–89
   of the ecological work 11, 20
   field data collection methods and reliability 92, 95
   in freshwater fish species guidance 137
      Allis Shad and Twaite Shad species guidance 145
      Sturgeon species guidance 149
      Whitefish/Vendace species guidance 141
   in Freshwater Pearl Mussel species guidance 113
   in Great Crested Newt species guidance 202
   identifying 66
   ignored in planning submissions 91, 208
   importance 30, 59, 66, 206
   to include in protected species guidance for planners in tables 95
   in invertebrate species guidance 181
   list in BS 42020 75
   missing in both of SA and SGN 55
   in Natterjack Toad species guidance 103

of NE SA advice given by NE 54
in Otter species guidance *123*
in protected plants, fungi and lichens species guidance *165*
SA's limited coverage of 60
unsuitable conditions 31
of using SGN format in practice 50
in Water Vole species guidance *118*
weather ignored as 91, *94*, *99*
in White-clawed Crayfish species guidance *134*
in wild bird species guidance *175*
local nature reserves (LNR) 89
Local Planning Authority (LPA) in species protection 2–3, 35
  biodiversity and 7–11, 18 (*see also under* biodiversity)
  bird advice and 52
  capacity, competence and LPAs 22–25
  Chelmsford District Council 12
  Chichester District Council 12
  differing expectations and requirements of 14
    Chelmsford District Council (CDC) *14*
    Chichester District Council (CC) *14*
    Huntingdonshire District Council (HDC) *14*
    Scarborough District Council (SBC) *14*
  Huntingdonshire District Council 12
  lists and checklists 24–25
  planning applications 2–3
  procedural steps 12–16
  Scarborough District Council 12
Local Record Centres (LRCs) 51, 74, *151*

Marsh, R. 213–214

Nash, D.J. *130*, 209
National Biodiversity Network (NBN) *151*, *159*
National Planning Policy Framework (NPPF) 5–6

Natterjack Toad, species guidance *101–105*
  field collection methods *102–103*
    calling males *102*
    qualitative data *102*
    quantitative *102*
    refugia use *102*
    spawn string count *102*
    toadlet counts *103*
    torch searches *102*
  impacts *104*
  site details *101*
  survey details *103–104*
  surveys objective *102*
Natural England (NE) guidance 3, 26, 27–32
  consulting 28
  planning permission refusal 29
  Planning Proposal for protected species (PPP) 27, 29–30
  Protected Species and Development (PSD) 27
  standing advice (SA) 30–32
  standing advice 28
  Standing Species Advice (SA) 27
  strands of 27
  suitable data 28–29
Natural England and DEFRA (NED) guidance 3, 32–46
  2022 32–46
  January 2022 suite of SA guidance in a nutshell 34–41
    pre-2022 and 2022 SA compared 39–40
    pre-2022 SAs 36–37
    site management and monitoring 38
    species/groups covered by 35
    PSD and PPP – 14 January 2022 33–34
  Scotland experience 41–46
    protected species advice 41–46
    NatureScot (2020a) PSSA versus January 2022 NE SA guidance 42–43
    new advisory guidance, comparison 44–45

Natural England's Ancient Woodland Inventory (NWI) *151*
Natural Environment and Rural Communities (NERC) *5*
Natural History Museum *2*
NatureScot *96*
NERC Act 2006 *16*
new 2022 Standing Advice *71–86*
    compensation for negative effects *81–82*
    data robustness, question of *79–80*
    enhancement *82*
    establishing facts *71–73*
    guidance, key features *76–77*
    January 2022 SA expectation on LPA planner *73–75*
    licence need *82–83*
    LPA planner needs *73*
    mitigation hierarchy sequence *80*
    NE 2022 SA guidance, issues with *84–85*
    planning decision by LPA *83–86*
    process initiator *75–78*
    proposed species development, assessing *79*
    qualification and licence *79*
    search for records and suitable habitats *78*
    steps *75–86*
North Lincs District Council *207*
Northern Ireland Strategic Planning Policy Statement (2015) *7*

Otter, species guidance *120–125*
    field collection methods *121–122*
        camera traps *122*
        DNA records *122*
        food remains *122*
        holt/den presence *122*
        presence of spraints under bridges or at weirs *121*
        presence of tracks on damp substrate *122*
        road kills *122*
        slides *122*
    impacts *124–125*
    mitigation measures *124*
    pre-2022 publication *224*
    site details *120*
    survey details *122–123*
    surveys objective *121*
Oxford, M. *3, 15–17, 21, 23, 31, 61, 83, 86*

Partnership for Biodiversity in Planning (PBP) *26, 46–61*
    2022 SA and 2020 SGN, comparison *56–59*
    individual species and group-specific best practice methods from *26*
    NE and WAC SGN species guidance notes *57–58*
    NE standing advice and SGN compared *54–56*
    SA and SGN at the species advisory level, comparison *59–61*
        badgers *59–60*
        bats *60–61*
        reptiles *60*
    SA and SGN guidance documents, comparison *55–56*
    Wildlife Assessment Check (WAC) *46–48*
    see also species guidance notes (SGN)
Peay, S. *131–134*
Planning Proposal for protected species (PPP) *29–30*
    preparing *27*
    PSD and PPP, 14 January 2022 *33–34*
planning system *5–25*
    reliable data in, importance of *12–16*
        assessing application, elements in, *15–16*
        differing expectations and requirements of *14*
    see also Association of Local Government Ecologists (ALGE), ecology and planning; Local Planning Authority (LPA) in species protection
preliminary ecological appraisal (PEA) *50, 88, 97*

preliminary roost assessment (PRA) *186*
professional competence *18*
protected designated sites in the UK,
    hierarchy of *8*
   sites of international importance *8*
   sites of national importance *8*
   sites of regional/local importance *8*
protected plants, species guidance
    159–167
   field collection methods *162–163*
     look-see method *163*
     systematic counts *164*
   impacts *166*
   monitoring *167*
   site details *159–160*
   survey details *164–165*
   surveys objective *160*
protected plants
   pre-2022 publication *228*
protected species
   delivering into the future *214–215*
   surveys, data from *206–215*
Protected Species and Development
    (PSD) *27–29*
   for local authorities *27*
   pre-2022 PSD advice *27*
   PSD and PPP – 14 January 2022 *33–34*
protected species guidance in tables
    94–205
   ancient trees and veteran trees (AVT)
     151–158
   ancient woodland *151–158*
   basic considerations *94–95*
   fungi *159–167*
   lichens *159–167*
   protected plants *159–167*
   see also under Badger; bats; freshwater
     fish; Freshwater Pearl Mussel;
     Great Crested Newt (GCN);
     Hazel Dormouse; invertebrates
     (IV); Natterjack Toad; Otter;
     Water Vole; White-clawed
     Crayfish; Whitefish/Vendace;
     wild birds
protected species standing advice
    (PSSA) *41*

protected species survey *87–205*
   avoidance *93*
   compensation *93*
   data age and sources *89*
   headings *87–93*
   licences and competencies *92–93*
   mitigation *93*
   monitoring *93*
   progressing *93*
   site name and geographic context *88*
   site status *89*
   species status *89*
   understanding impacts *93*
   see also field data collection methods
     and reliability

Rampling, E.E. *208, 210, 214*
reptiles, species guidance *60, 125–130*
   field collection methods *127*
     directed visual transects *127*
     refugia searching *127*
     skin sloughing *127*
   impacts *129–130*
   pre-2022 publication *225*
   site details *125–126*
   survey details *127–128*
   survey timing
     Adder *128*
     Common Lizard *127*
     Grass Snake *128*
     Sand Lizard *127*
     Slow Worm *128*
     Smooth Snake *128*
   surveys objective *126*
Robertson, M. *74, 208, 214*
roost characterisation survey (RCS) *188*

Sand Lizard *125–126*
Scarborough District Council (SBC)
    12, 14
   differing expectations and
     requirements of *14*
Scotland, protected species advice
   in *41–46*
Scottish Planning Policy *7*
Sewell, D. *200–202*

site of special scientific interest (SSSI) 28, *136*
Snell, L. 16, 31, 86
Special Areas of Conservation (SAC) 88, *136*
species guidance notes (SGN) 48–54
  and badgers 50
  bird advice and the LPA planner 52
  bird risk and survey needs (BSG 2021) 53
  birds and 50
  birds and bird survey guidelines 51–54
  contents and structure of 49
  format in practice 49–50
  reptiles and 50
species-specific methods (SSM) *172*
standing advice (SA), Natural England 30–32
  high-level categories in 31
static recording/automated survey *189*
Sturgeon, species guidance *147–150*
  field collection methods *149*
  impacts *150*
  monitoring *150*
  site details *147–148*
  survey details *149–150*
  surveys objective *148–149*

Thompson, S. 213
Tyldesley, D. 17–18

Vantage point surveys (VPS) *169, 171*
Vendace *139–143*

Water Vole, species guidance *115–119*
  field collection methods *116–117*
  impacts *118*
  pre-2022 publication *226*
  site details *115*
  survey details *117–118*
  surveys objective *116*
Waterfowl Counts (WFC) *169, 171*
White-Clawed Crayfish, species guidance *131–135*
  field collection methods *132–133*
    fixed-area sampling *132*
    night viewing *133*
    standard method *132*
    trapping (baited) *132*
    trapping (unbaited) *132*
  impacts *134*
  pre-2022 publication *227*
  site details *131*
  survey details *133*
  surveys objective *131–132*
Whitefish/Vendace, species guidance *139–143*
  field collection methods *140–141*
  impacts *141–142*
  site details *139–140*
  survey details *141*
  surveys objective *140*
wild birds, species guidance *167–178*
  field collection methods *169–172*
    breeding bird survey *169*
    Common Bird Census Breeding bird survey (CBC) *170*
    species-specific methods (SSM) *169, 172*
    vantage point surveys (VPS) *169, 171*
    waterfowl counts (WFC) *169, 171*
    windfarm assessments (WFA) *169, 172*
    winter bird survey (WBS) *169, 170*
  impacts *176–177*
  monitoring *177*
  site details *167–168*
  survey details *174–175*
  survey timing *173*
  surveys objective *168–169*
Wildlife and Countryside Act 1981 6, *136, 161*
Wildlife Assessment Check (WAC) 12, 26, 46–48
  web-based screening check 47
windfarm assessments (WFA) *169, 172*
windfarms 28, 50, *191, 193–194, 195, 196*
Winter Bird Survey (WBS) *169–170*

Zu Ermgassen, S. 82, 210, 212, 214

Printed in the USA
CPSIA information can be obtained
at www.ICGtesting.com
LVHW081752021124
795283LV00002B/3